# International Symposium on Bioinformatics, Chemometrics and Metabolomics 2016



Journal of Physics: Conference Series Volume 835

Bogor, Indonesia
18 October 2016

ISBN: 978-1-5108-4077-5
ISSN: 1742-6588

**Printed from e-media with permission by:**

Curran Associates, Inc.
57 Morehouse Lane
Red Hook, NY 12571

**Some format issues inherent in the e-media version may also appear in this print version.**



Printed by Curran Associates, Inc. (2017)

For permission requests, please contact the Institute of Physics
at the address below.

Institute of Physics
Dirac House, Temple Back
Bristol BS1 6BE UK

Phone: 44 1 17 929 7481
Fax: 44 1 17 920 0979

techtracking@iop.org

**Additional copies of this publication are available from:**

Curran Associates, Inc.
57 Morehouse Lane
Red Hook, NY 12571 USA
Phone: 845-758-0400
Fax: 845-758-2633
Email: curran@proceedings.com
Web: www.proceedings.com

# Table of contents

## Volume 835

**International Symposium on Bioinformatics, Chemometrics and Metabolomics**

**18 October 2016, Bogor, Indonesia**

**Accepted papers received: 24 April 2017**
**Published online: 17 April 2017**

### Preface



### Papers

# Preface

The International Symposium on Bioinformatics, Chemometrics and Metabolomics (ISBCM) 2016 was organized by Tropical Biopharmaca Research Center, Metabolomics Research Cluster-Directorate of Research and Innovation and Working Group on Bioinformatics FMIPA Bogor Agricultural University (IPB), supported by Indonesian Chemical Society, The Indonesian Society for Biochemistry and Molecular Biology and Indonesian Society for Bioinformatics and Biodiversity. This event aims at sharing knowledge and expertise, as well as building network and collaborations among scientists working on Bioinformatics, Chemometrics and Metabolomics (BCM), which are vibrant multidisciplined fields with positive trend in this decade.

Metabolomics is an "advanced research" approach used to improve our understanding of metabolic systems and biochemistry compositions of plants and animals. Chemometrics is a science of how to extract information from chemical systems that can be applied to solve both descriptive and predictive problems in experimental life sciences. Bioinformatics refers to conceptualizing biology in terms of macro molecule (from in chemistry and physics points of views) then applying computational approaches from informatics (applied mathematics, computer science and statistics) in order to understand and manage molecular information in its macro scale. This year's theme is ***The emerging era of bioinformatics, chemometrics, and metabolomics in agriculture, food and health: current success, future potentials and challenges***. This conference embraces all aspects of BCM applications for solving real problems in food, health and agriculture. High quality, original papers are solicited in all areas of Bioinformatics, Chemometrics and Metabolomics.

ISBCM2016 was organized as a scientific symposium, which covers a keynote speech, some invited talks, and some contributed presentations and posters. At this moment, the meeting brought together about 100 participants who are composed of many active researchers, government officers, and entrepreneurs. There were 3 keynotes and 6 invited speakers sharing their expertise coming from Australia, Indonesia, Malaysia and Singapore. There were 31 oral presentations and 38 poster presentations coming from Indonesia and overseas. In conjunction with this symposium we also held a workshop on application of chemometrics for quality control of medicinal plant.

In this occasion, we were honored to have our distinguished keynote speakers Prof. Dr. Mohd Zaki Salleh (Universiti Teklogsi MARA), Dr. Sanjay Swarup (National University of Singapore), and Dr. Beben Benyamin (University of Queensland, Australia) and invited speakers Prof. Dr. Abdul Rahman (Gadjah Mada University, Indonesia), Prof. Dr. Teh Lay Kek (Universiti Teknologi MARA, Malaysia), Prof. Dr. Yaya Rukayadi (Universiti Putra Malaysia), Dr. Nancy Dewi Juliana (Bogor Agricultural University, Indonesia), Dr. Farit M Afendi (Bogor Agricultural University, Indonesia), and Dr. Awang Maharijaya (Bogor Agricultural University, Indonesia). We also thank all participants for their contributions especially those who submit their full papers to the editorial board.

Finally, we also thank our reviewers for maintaining a high quality standard of the accepted manuscripts for publication.

Guest Editors

Dr. Eng. Wisnu Ananta Kusuma, ST., MT. (Bogor Agricultural University, Indonesia)
Prof. Dr. Abdul Rohman M.Si., Apt (Gadjah Mada University, Indonesia)
Dr. Sastia Prama Putri (Osaka University, Japan)

**Figure.** Speakers and participants of ISBCM 2016

IOP Conf. Series: Journal of Physics: Conf. Series **835** (2017) 011002 doi:10.1088/1742-6596/835/1/011002

# Peer review statement

All papers published in this volume of *Journal of Physics: Conference Series* have been peer reviewed through processes administered by the proceedings Editors. Reviews were conducted by expert referees to the professional and scientific standards expected of a proceedings journal published by IOP Publishing.

IOP Conf. Series: Journal of Physics: Conf. Series **835** (2017) 012001 doi:10.1088/1742-6596/835/1/012001

# Detection of protein complex from protein-protein interaction network using Markov clustering

**P J Ochieng**[1,2], **W A Kusuma**[1,2,3*] **and T Haryanto**[1,3]

[1]Department of Computer Science, Faculty of Mathematics & Natural Sciences, Bogor Agricultural University Dramaga, Bogor 16680, Bogor, Indonesia
[2]Kenyatta National Hospital, P O BOX 20723-00202, Upper hill, Nairobi, Kenya
[3]TropicalBiopharmaca Research Center, Bogor Agricultural University, Jl. Taman Kencana No. 3, Bogor 16128, Indonesia

E-mail: ananta@apps.ipb.ac.id

**Abstract.** Detection of complexes, or groups of functionally related proteins, is an importantchallenge while analysing biological networks. However, existing algorithms to identify protein complexes are insufficient whenapplied to dense networks of experimentally derived interaction data. Therefore, we introduced a graph clustering method based on Markov clustering algorithm to identify protein complex within highly interconnected protein-protein interaction networks. Protein-protein interaction network was first constructed to develop geometrical network, the network was then partitioned using Markov clustering to detect protein complexes. The interest of the proposed method was illustrated by its application to Human Proteins associated to type II diabetes mellitus. Flow simulation of MCL algorithm was initially performed and topological properties of the resultant network were analysed for detection of the protein complex. The results indicated the proposed method successfully detect an overall of 34 complexes with 11 complexes consisting of overlapping modules and 20 non-overlapping modules. The major complex consisted of 102 proteins and 521 interactions with cluster modularity and density of 0.745 and 0.101 respectively. The comparison analysis revealed MCL outperform AP, MCODE and SCPS algorithms withhigh clustering coefficient (0.751) network density and modularity index (0.630).This demonstrated MCL was the most reliable and efficient graph clustering algorithm for detection of protein complexes from PPI networks.

## 1. Introduction

In genomics studies, large-scale experiments often produce huge data sets of protein-protein interactions making it difficult to visualize and analyze the information contained in these data [1][2]. Therefore, application of computational methods can alleviate a lot of problems in this regard. Thus a general trend is to represent the interactions as a network/graph and to apply suitable graph algorithms to extract necessary information. In the post-genomic era, the most important issues are to find protein complexes from the protein-protein interaction (PPI) networks. Identification of those complex can help us to predict the functions of proteins [3], and they can also play an important role in understanding and explain certain biological processes.

* Corresponding author

The results obtained from different technologies for detection of high-throughput protein-protein interactions such as yeast two-hybrid assay (Y2H) and mass spectrometry of purified complexes say tandem affinity purification (TAP) and high-throughput mass spectrometric protein complex identification (HMS-PCI) show some variations[4][5]. For instance, the common PPI between the two different mass-spectrometry approaches stands at 1,728 pairs, which correspond to 27.5% of PPI detected by TAP and 19.2% of PPI detected by HMS-PCI. This implies that many variations in experimentally determined interaction might be false positive are incomplete. Hence, generation of protein complexes based on interaction networks of separate or combined data sets is helpful because the interactions that are involved in complexes are likely to be true.

Already a number of approaches have been proposed for detection of protein complexes in PPI networks[6]. The sequential constructive method of makes use of the concepts of clustering coefficient and k-core graphs [7] [8].

Another approach described in [9] use hierarchical clustering. However, they introduced the concept of secondary distances instead of considering the path length as the distance between a pair of proteins because of the fact that such distances among proteins are constrained and often cause distance ties[10]. The approach of [11] starts by composing an initial random clustering and then iteratively moving one node from one cluster to another in a randomized fashion to improve the clustering's cost [12]. Once the clusters are generated, they are filtered based on cluster size, density and functional homogeneity keeping in mind the criteria of the known biological complexes. Another approach related to analyzing protein complexes is super-paramagnetic clustering [13] [14]. Furthermore, module based approach such as the Molecular Complex Detection (MCODE), Affinity Propagation (AP) and Super Paramagnetic Clustering (SPC) have been proposed recently to detect protein complexes [15, 16,17].

Recently, AP have been proposed in the genomic study of complex proteins in PPI network since it identifies exemplars within the dataset by exchanging real-valued messages between all data points are then grouped with their most representative exemplar to give the final set of clusters [18][19]. Furthermore, Affinity Propagation (AP) have been applied gene identification from putative exons using microarray data, and have proven to be faster and more accurate than other clustering methods like K-Centers clustering algorithm. However, studies have demonstrated the algorithm function similar to Vertex Substitution Heuristic (VSH) algorithm which is often less reliable for protein complex detection [20, 21, 22].Recently studies have suggested that clusters or locally dense regions of an interaction network might represent protein complexes [23]. However, the term "locally dense region" implies a very flexible concept. Some well-known clustering methods are k-core, k-block, k-plex and n-clan clustering where k-core is a maximal subgraph such that each node in the subgraph has at least degree k, however strategies are usually based on the number of node degrees or the number/length of paths between two nodes within the cluster which make it challenging for detection of complex protein in large PPI networks [24, 25, 26].

Therefore, in this paper we proposed Markov Clustering (MCL) to partition the PPI network to facilitate detection of complex proteins. Thus we assume that the interaction network is an undirected simple graph where a graph is undirected if its edges are not directed and a graph is simple if it has no parallel edge or self-loop. We also initially perform MCL parameter optimization before identification of protein complexes from PPI network based on the network modularity and density from the MCL cluster granules.

## 2. Material and Methods

### *2.1. Data preparation*

In this study, 2053 protein datasets associated with type 2 diabetes diseases obtained from Human Reactom Protein Database (HRPD) [27]. The dataset was pre-processed to eliminate any artefact and check for missing data.

IOP Conf. Series: Journal of Physics: Conf. Series **835** (2017) 012001 doi:10.1088/1742-6596/835/1/012001

### *2.2. Construction of PPI Network*

Protein-protein interaction network (graph) was initially built to link protein pair that share similar biological function. Where G represented the graph (network), and the vertex (V) be the node (protein) and the edge (E) be the interaction (protein interaction). Therefore, we represented graph G = (V,E) where mathematically it consist of finite nonempty sets V of nodes(proteins) where, a set $E \subseteq V \times V$ of edges consisting of unordered pairs of vertices (proteins) with connecting edges ( E). To extract complex protein network we build a subgraph (network) in this case we let subgraph (H) be H = (V H, E H) of protein-protein G = (V, E) if $VH \subseteq V$ and $EH \subseteq E$. We also say that G is a supergraph of H. Given a subset of the vertices $V' \subseteq V$. In this case, the induced subgraph G′ = (V′, E′) consists exactly of all the edges (protein interaction) present in G between vertices in V′. In this case, for all $v_i, v_j \in V'$, $(v_i,v_j) \in E' \Longleftrightarrow (v_i,v_j) \in E$. In other words, two nodes are adjacent in G′ (protein complex) if and only if they are adjacent in G (protein network). From the constructed network, it is easy to identify the neighbour distribution and connectivity, therefore the next step is to partition the network via MCL to detect the sub-network (complexes) with highest interaction and degree distribution.

### *2.3. Network partioning via MCL algorithm*

To partition the PPI network we subject the network to Markov Clustering (MCL). In this case, we let G be a graph on n nodes and $M = M_G$ be the associated matrix. We then denote graph G associated with Markov matrixby $\mathcal{T}_G$ which we formally defined by letting its q[th] column be the q[th] column of M normalized. Since our graph is undirected, we let $M = M_G$ be the associated matrix and $\mathcal{T} = \mathcal{T}_G$ be the associated Markov matrix. Again, to this end, we used ***d*** to denote the diagonal matrix that has diagonal entries the column M, thus$d_{kk} = \sum_i M_{ik}$, and $d_{ij} = 0$; $i \neq j$ while $\mathcal{T}_G$is defined by (1).

$$\mathcal{T}_G = M_{G^{d-1}} \tag{1}$$

Then the number of positive, negative and zero eigenvalues are the same for T and M. In this case the matrix $M$ $\mathbb{R}^{k\times l}$, $M \geq 0$, and a real nonnegative number r, the resultant matrix from rescaling each of the columns of M with power coefficient r is represented by $\Gamma_r M$, where Γr is the inflationoperator with power coefficient r. Thus, theoretically, the action of $\Gamma_r$: $\mathbb{R}^{k\times l} \rightarrow \mathbb{R}^{k\times l}$ is defined by(2).

$$(\Gamma_r M)_{pq} = \left(M_{pq}\right)^r / \sum_{i=1}^{k} \left(M_{pq}\right)^r \tag{2}$$

In this case, we tune the inflation parameter (r) in (2) to generate the optimal number of clusters. The aim of tuning the inflation operator is to make it stronger, to enhance cluster granularity or tightness. Therefore, by setting maximum iteration for expansion and inflation we enhance the graph partitioning into different segments until equilibrium is achieved hence we interpret the collections of segmentsas protein complexes. Below isrepresentation of the proposed MCL algorithm (Figure 1).

**MCL Algorithm**

```
  G is a graph
  add loops to G
  set Γ to some value      # affects granularity
  set M_1 to be the matrix of random walks on G

  while (change) {
M_2 = M_1 * M_1                # expansion
M_1 = Γ(M_2)                   # inflation
    change = difference(M_1, M_2)
  }

  set CLUSTERING as the components of M_1
```

**Figure 1**. MC algorithm

### *2.4. Parameterization*
To optimize the clustering performance of the proposed MCL we initially perform flow simulation by setting the inflation value (r) within a range of 1.4 to 2.5 power coefficient this was according to [28]. The ultimate goal was to determine the optimal inflation value for maximum cluster granularity in other words efficient network partioning. Thereby, in this study flow simulation acts as preliminary MCL optimization for accurate detection of protein complexes.

### *2.5. Detection of protein complexes from MCL clusters*
In the current study protein complexes from the network were calculate based modularly score and density of the protein community (cliques) cluster granules generated by MCL. Thereby we used the formulation in (2.5.1) and (2.5.2) to determine the significant protein complex.

*2.5.1 Modularity score.* To detect protein complexes from the MCL networks we computed the modularity for each network using global modularity calculation according to [29] and [30]. This measure provides a composite modularityscore across all clusters and is defined as:

$$\text{Modularity} = \sum_{c \in C}\left[\frac{E_{cIn}}{E_{total}} - \left(\frac{2E_{cIn} + E_{cOut}}{2E_{total}}\right)\right] \tag{3}$$

where, for each cluster c in the set of all clusters C, $E_{cIn}$, $E_{cOut}$ and $E_{total}$ represent the number of edges within the cluster, the number of edges leading out of the cluster, and the total edges in the network, respectively. We note that while the global modularity score only considers clusters that are contained within the main graph component, in practice this does not significantly affect the results because few or no clusters in the networks we consider them isolated from the main component. In some case we calculate local modularity for each cluster using (4).

$$C_{mod} = E_{in}/E_{out} \tag{4}$$

*2.5.2 Density.* Calculation of cluster densities plays an important role in identifying the major complexes in the partitioned PPI Network. Here be considered density $d_k$of any cluster k as the ratio of the number of edges present in the cluster ($|E_k|$) and the maximum possible number of edges in the cluster ($|E_k|_{max}$) and is represented by (5).

$$d_k = \frac{|E_k|}{|E_k|_{max}} = \frac{2 \times |E_k|}{|N_k| \times (|N_k| - 1)} \tag{5}$$

Thus, $|N_k|$ is the size of the cluster, i. e. the number of nodes in the cluster however it should be noted that density of a given cluster is a real number ranging from 0 to 1.

## 3. Results
The prime goal of the current study was to efficiently detect protein complexes from PPI network via MCL partitioning and network topological calculation, therefore we initially generated the PPI network (3.1) then perform the flow simulation (3.2) to determine optimal parameters values for MCL this was then followed by analysis of overall quality of complex (3.3) finally evaluation of MCL performance results (3.4).

### *3.1. Original PPI network*
The unweighted PPI network was built 2254 protein pairs of proteins. The resultant network (Figure 2) consist of 482 nodes (proteins) with 2054 edges (interactions) with network density = 0.08 and

average neighbour connectivity= 8.51 demonstrating that there is strong transitive relationship (interaction) between protein pairs in the network.

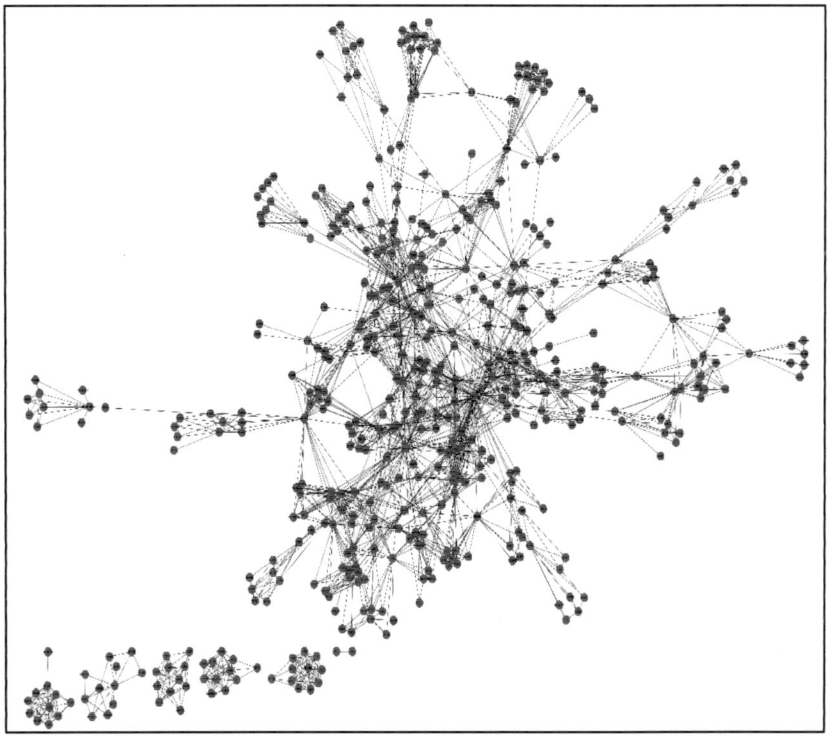

**Figure 2.** Original unweighted PPI network comprising of 482 proteins and 2054 interaction generated from protein interaction pairs associated with type 2 diabetes.

*3.2. Flow simulation*

In this study we initially perform flow simulation to search optimal inflation value for maximum cluster granularity, therefore we set the inflation value within a range of 1.4 to 2.5 simply because the inflation parameter is responsible for both strengthening and weakening the current (strengthen the already strong currents and weaken the already weak currents). Furthermore, by setting the inflation range we are able to obtain the optimal inflation value with high strengthening or weakening property in the end influences the granularity of clusters. From our simulation adjusting the inflation influence the average size of cluster and the maximum size of cluster as this had the over impact on the number of clusters generated from the original PPI network (Figure 3). Flow simulation clearly demonstrate setting inflation parameter at R =1.4 resulted in 13 protein complexes (clusters granules) likewise when we set the maximum value R= 2.5 number of the identified protein complexes is extremely high (49 clusters) leading to over-representation due to high strengthening of current this result in high cluster granularity and this may not reflect the true protein complexes within the PPI network.

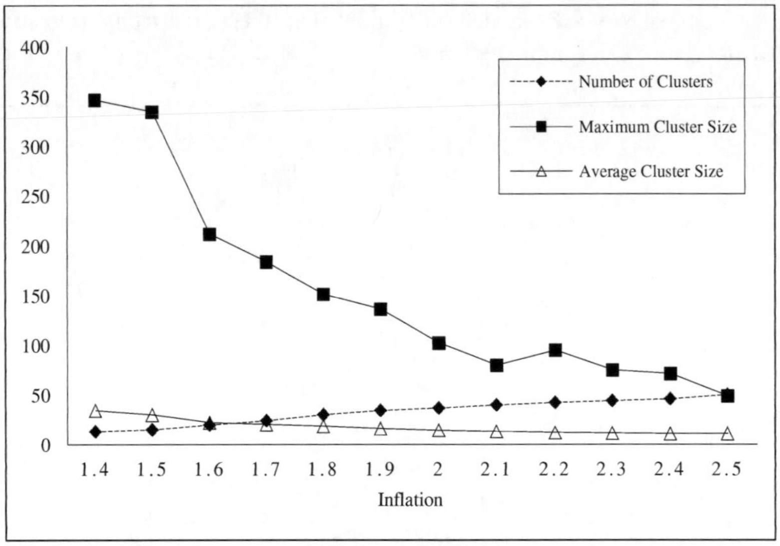


**Figure 3.** Effect of inflation adjustment on average cluster size, number of cluster granules and maximum cluster size.

This implies by setting inflation R=1.4 the resultant output is a set of nested overlapping clustering's on the other hand extremely high inflation value leads to high cluster granularity resulting to under-representation of the true clusters. Interestingly, it can also be evidence from the result as the number of inflation increase gradually the number of detected protein complexes also increases. This is very satisfactory, as one expects clusters at different levels of granularity to be related to each other. It holds that when inflation parameter is set sufficiently low value, the network will be clustered into a single cluster. However, when the dealing with large networks where the diameter is extremely large, the algorithm is likely to experience time complexity since random network are built on distinctly different probability distributions several expansion steps at optimal inflation parameter is required equalize the matrix. Thus in this analysis the optimal inflation values obtained at value 1.9 which gave 34 complexes (clusters) with a maximum cluster size of 136 (Figure3 ) and network density of 0.14 as well as the modularity score of 0.61 (Figure 4).

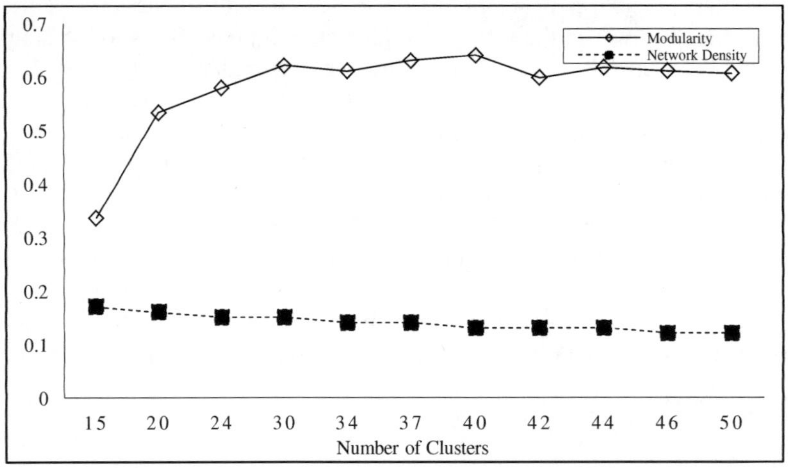


**Figure 4.** Effect of change in number of clusters to network modularity and density.

We further assess the effect inflation value on modularity and network density (Figure 5), from the simulation we discovered that an modularity score increase with rise in inflation value whereas the network density decrease as we increase the inflation values, in network perspective cluster modularity and density represents the level of connectivity within a group of nodes relative to the group's connections to the rest of the network. Previously (2.5.1) we defined modularity as the ratio of the number of edges between nodes in a cluster (in-degree, $E_{in}$) to the number of edges between members of the cluster and any neighbors not designated as members of the cluster (out-degree, $E_{out}$) therefore by performing the flow simulation. In our case, high modularity score indicated that a cluster is very isolated from the rest of the network (Figure 4).

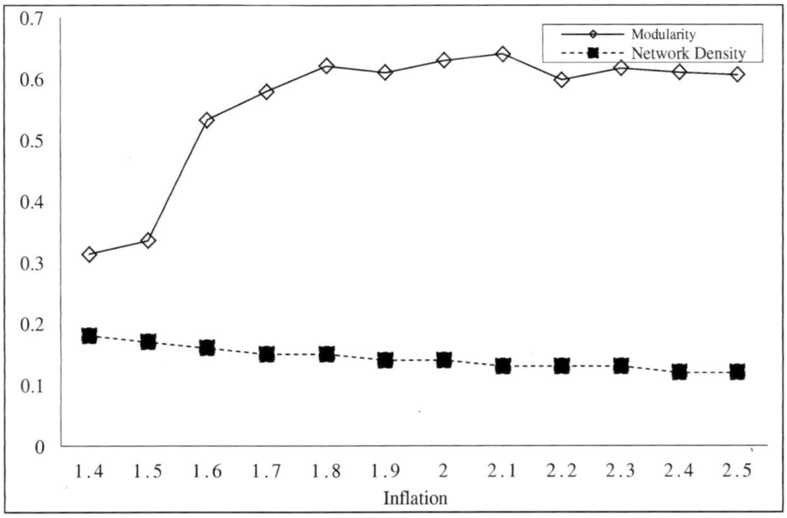


**Figure 5.** Effect of adjusting inflation parameter on PPI network modularity and density

Performing flow simulation enable us choose the optimal inflation value for MCL algorithm for maximum networktherefore set optimal inflation value MCL generate 34 clusters modules (protein complex) with the major complex consisting of 103 proteins and 521 interactions (Figure 6) likewise smallest protein complex consisted of 3 protein and 4 interactions (Figure 7). Generally, average clusters modularity was 0.63 with the clustering coefficient of 0.742 and density of 0.077. This demonstrated that majority of the clusters highly connected probably due to optimal inflation value, furthermore the majority of complex identified had 100% shortest path between nodes within cluster this indicated strong intra-clusters. This played an important role discovery of proteins sharing similar biological function.

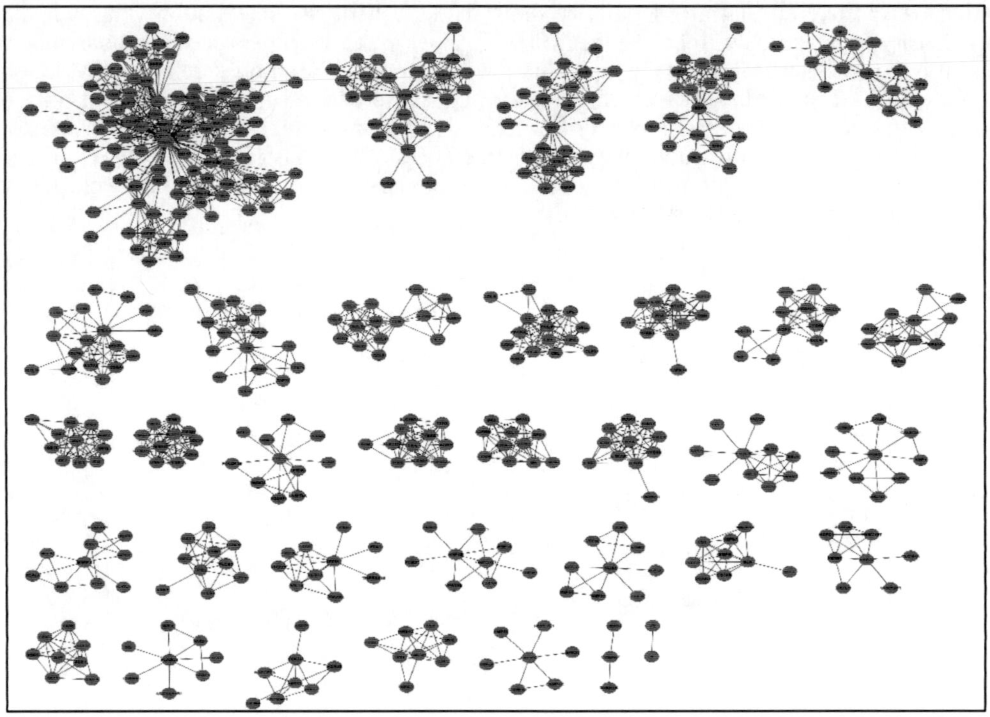

**Figure 6.** MCL clusters oforiginal PPI network (3.1). 34 cluster modules are generated from PPI network with five major complexes consisting of >20 protein interaction.

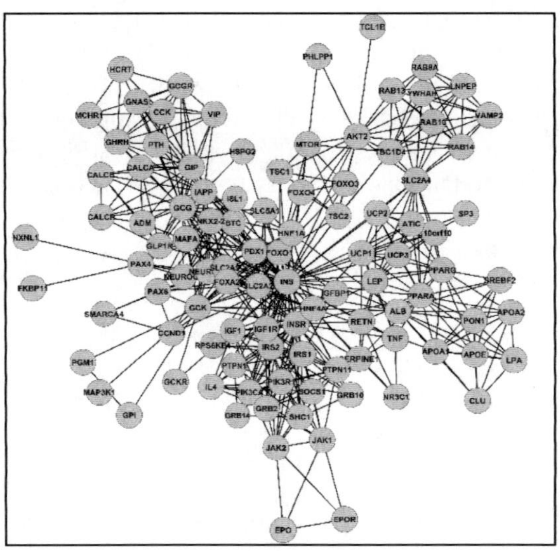


**Figure 7.** Major complex extracted from MCL clusters. It is the subnetwork with the highest number interaction (edge) and node (vertex) connectivity.

*3.3. Overall quality of the detected complexes*

We discovered that 78% of 2054 interactions (involving a total of 482 proteins) occurred between pairs from the original PPI network (Figure 1) at least shared a common function. Though the majority

IOP Conf. Series: Journal of Physics: Conf. Series **835** (2017) 012001 doi:10.1088/1742-6596/835/1/012001

of the interactions are between similar function protein pairs, there are many instances of interactions betweenproteins of different functions (overlapping modules). Therefore, it was reasonable to assume that interactions that forms the protein complex arebetween similar function protein pairs. In other words, it can be said that the quality of detected complex is goodif it contains mostly similar function proteins. Thus, to check the quality of the detected complexes, we estimate therelative amount of interactions that are between similar function protein pairs out of intra-complex interactionsfor complexes using equation below (6).

$$RA = \frac{\bigcup_{i=0}^{n} SFI_i}{\bigcup_{i=0}^{n} AI_i} \tag{6}$$

where, $n$ is the number of complexes of size $\geq 3$ in a set, $SFI_i$ is the number of interactions of cluster $i$, that are between protein pairs of identical functional class, and $AI_i$is the number of all interactions in cluster $i$ of the corresponding set Figure. 6 shows the relation between $RA$ and $d_{in}$. There is a sudden rise from $din$ = 0.101 to $din$ = 0.268. Thus the complexes having density 0.101 or less have high statistical significance. When complexes are generated using $d_{in}$= 0.268, many of the complexes consist of more than three proteins and two interactions on the other hand when $d_{in}$= 0.101 is used, the aforementioned type of 3-protein complexes are excluded. Therefore, it may be concluded that many of the interactions contained in 3-protein complexes of density 0.268 are not interactions between similar function proteins pairs since the functions of all the proteins involved in the network or complexes are not yet known.It is noticeable that the percentage of interactions between similar function protein pairs is higher in high-densitycomplexes and this percentage might increase if the functions of all the proteins were known(Figure 8). Hence it can be concluded that the interactions that form high-density clusters in PPI networks represent functional complexesand can be considered as true interactions with high chances.

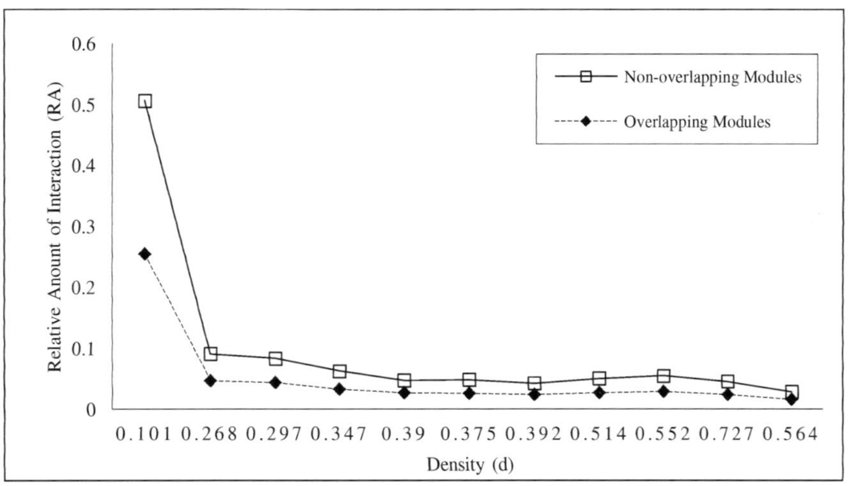


**Figure 8.** Relative amount of interactions involving non-overlapping and overlapping modules in MCL 34 clusters (protein complexes) against corresponding densityvalues

### *3.4. Evaluation of MCL performance*

To assess the robustness of MCL in the detection of protein complexes in the PPI network wefurther compared MCL performance with MCODE, SCPS, and AP using similar protein dataset (Figure 9). Thus the included the adjustable parameters for the three tested algorithms (MCODE, SCPS, and AP)

and evaluated across a wide spectrum of their settings. The performance of all algorithms initially assessed in terms of density, modularity, and clustering coefficient of identified clusters with respect to a number of detected clusters (protein complexes). MCL and SCPS algorithm generated high modularity (0.630 and 0.652 respectively) and cluster coefficient (0.751 and 0.731 respectively)with the both algorithm generating clusters with similar density (0.014). The three algorithms also tend to produce a much greater variation in the total number of clusters identified across their parameter settings, while still producing clusters of lower modularity; this is particularly striking for MCODE (0.304). Furthermore, cluster granule produced by both AP and MCODE were of low density implying that there was less interaction with protein in each of those complexes this explain why both MCODE and AP had low clustering coefficient (0.573 and 0.595 respectively).

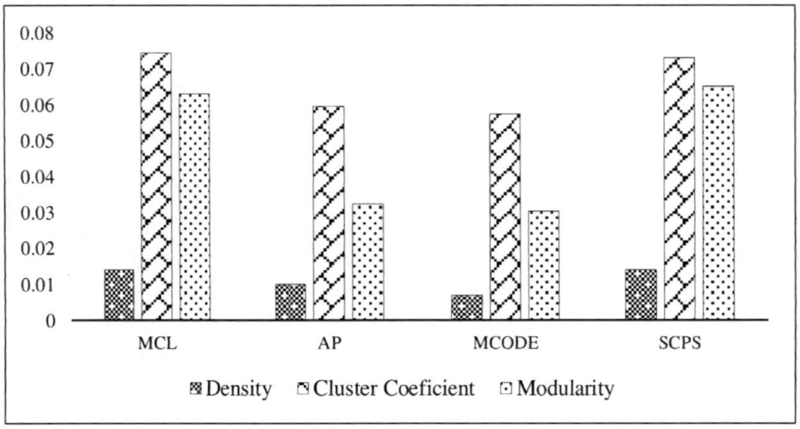


**Figure 9.** Comparison of clustering performance of MCL with AP, MCODE and SCPS algorithm based on cluster coefficients, modularity and cluster density.

We compared the performance of the proposed MCL with other three algorithms (MCODE, SCPS, and AP) in terms of the number of identified complexes (clusters), average cluster size and minimum cluster size (Table 1). From the analysis it was quite evidenced that MCL outperform MCODE, SCPS and AP algorithm, the MCL successfully detect 34 complexes (clusters) with the largest component comprise of 136 proteins with 711 interactions. Again, it is evidence that the MCL algorithm is able to archive strong cluster-wise separation with an average cluster size of 14.176. In addition, MCL present the highest inter-cluster range with 99 the algorithm also gives quite impressive modularity index (0.63) meaning that the identified components consist of highly interacting proteins. MCODE performs quite better than SCPS and AP by yielding 29 clusters with average cluster size being 11.103. Despite MCODE yielding better inter-cluster range 72, the algorithm record the lowest modularity index (0.304) may probably explain the weak cluster-wise separation when compared to MCL algorithm. On the other hand, SCPS algorithm only yields 26 clusters with highest an average cluster size of 18.538 and modularity index (0.64). Nevertheless, the AP performs the worst when subjected to the unweighted graph, in this analysis we had to set arbitrarily preference weight of -1.0 to facilitate the clustering. Although AP gave, the highest number of clusters (60) there was biases due to cluster over-representation. Therefore, it still holds that MCL outperform AP when clustering unweighted graphs.

**Table 1.** Evaluation of clustering performance by optimal parametrization.

| | Parameter | Optimization values | Number Clusters | Average Cluster size | Maximum Size | Minimum Size |
|---|---|---|---|---|---|---|
| **MCL** | Inflation (r)<br>Maximum iteration<br>Cut-off | 1.9<br>100<br>1.0 | 34 | 14.176 | 103 | 3 |
| **SCPS** | Epsilon (ε) | -0.1 | 26 | 18.538 | 63 | 4 |
| **MCODE** | Depth<br>Haircut<br>fluff | 100<br>TRUE<br>FALSE | 29 | 11.103 | 75 | 3 |
| **AP** | Lambda (λ)<br>Edge cut-off<br>Arbitrarily weight<br>preference | 0.5<br><br>1<br>-0.1 | 60 | 8.033 | 110 | 1 |

## 4. Discussion

Detection of protein complexes using existing clustering algorithm poses a great the challenges in the detection of significant biological interactions contained in the PPI networks. The MCL algorithm is a fast and scalable unsupervisedclustering algorithm. It is one of the most widely usedalgorithms and is it is based on simulating stochasticflows in networks. The MCL algorithm can detect cluster structures in graphs by taking advantage of a mathematical bootstrapping procedure. The process is tryingto perform random walks through a graph and deterministically compute their probabilities to find the bestpaths. It does so by using stochastic Markov matrices.The algorithm works by alternating the inflation parameter, which iteratively calculate the set of transitionprobabilities. The inflation operator implements a stochastic matrix transformation to emphasize larger probabilities and deemphasize smaller ones [31].Affinity propagation is an unsupervised algorithm andthus the number of clusters are automatically calculated.The idea behind this algorithm is to find sub-paths,which allow easy message exchanges between nodes. Ittakes as input a similarity matrix, which keeps the distances between all possible pairs of data points whereasit initially considers all data points as potential "exemplars". In later steps, real-valued messages are exchangedbetween the nodes until a set of exemplars and corresponding clusters emerges with high quality. The mainparameter of this algorithm is the 'preference', whichcontrols how many data points are selected asexemplars [32].Spectral clustering tries to detect clusters in a graph, wherenodes are connected with highly-similarity. The algorithm also tries to find connections between such areasthat should be weak, constituted by edges of low similarity. The aim is to identify highly connected clustersand, at a later stage, filter the inter edges within thecluster. The only parameter required is the user-definednumber of clusters [33]. In the current study, we used MCL to partition PPI network in order to detect protein complexes. In MCL clustering occasionally there is overlap at several levels of cluster granularity this always corresponds network asymmetry for undirected graphs which are always characterized by overlapping modules. We realised global cluster modularity plays an significant role in the detection of major complexes in partitioned PPI network since by computing the global modularity score from a composite of the local modularity across all clusters we accounts for edges inside each cluster, edges connecting each cluster to the rest of the network, and the total number of edges in the network. When evaluated over a range of parameters, we realised the proposed MCL algorithm produce cluster granularity from PPI network with higher modularity than other methods (AP, MCODE, and SCPS) this is because MCL perform random walk within the PPI network taking into account the expansion and inflation this make it more efficient for detection of protein complexes than other methods, with the exception of a single setting of epsilon value for SCPS, which result in higher modularity however with less number of identified clusters. On other hand AP and MCODE produced large number clusters however with low modularity and density due to less interaction within the clusters. Furthermore, AP yields a high number of complexes, the algorithm seems to be the most sensitive to the unweighted graph thus making it less reliable for detection of protein complexes. Therefore, the proposed MCL algorithm can be considered as the most stable one and performs best when partitioning both unweighted and weighted graphs.

## 5. Conclusion

In this study it's clearly evidenced that the proposed MCL algorithm is highly tuneable graph-clustering algorithm for protein complexes detection in dense and highly interconnected PPI networks.Performing flow simulation enable us obtained optimal inflation value for MCL enabling the algorithm to generate maximum cluster granularity. This facilitate detection proteins that share common biological functional within the clusters.Moreover, by computing cluster modularity and density we identified major protein complex with high protein interaction from partitioned the PPI networks. Further empirical comparison indicates MCL to be robust and reliable for detection of protein complexes as opposed to AP, SCPS and MCODE algorithm especially when we set optimal parameters of each algorithm although, we discovered under most conditions, AP and MCL outperform MCODE and SCPS. It still holds MCL is remarkably robust to variations in the choice of parameters, whereas the other algorithms require appropriate tuning in order to yield relevant results as we witness with AP one need to set arbitrary preference weight to facilitate clustering. This confirmed the general superiority of MCL over the three tested algorithms making suitable graph partioning algorithm for detection of complex proteins in the PPI network.

## References

[1] Bader G, Hogue C 2003 *BMC Bioinf.* **4** 27
[2] Schwikowski B, Utez P and Fields S 2000 *Nat. Biotechnol.* **18** 1257-61
[3] Ho Y, Gruhler A, Heilbut A, Bader G D, Moore L, Adams S L, Millar A, Taylor P, Bennett K, Boutilier K, et al 2002 *Nature* **415** 180-3
[4] Peng J, Elias, Thoreen C, Licklider L and Gygi P 2002 *J Proteome Res.* **10** 1021
[5] Uetz P, Giot L, Cagney G, Mansfield TA and Judson RS, et al 2000 *Nature* **403** 623–7.
[6] Edachery J, Sen A and Brandenburg J 1999 *Lect Notes Comput Sc.* **1731** 98-106
[7] Matula D 1972 *SIAM J. Appl. Math.* **22** 459-80
[8] Arnau V, Mars S and Marin I 2005 *Bioinformatics* **21** 364-78
[9] King A D, Pržuli N and Jurisica I 2004 *Bioinformatics* **20** 3013-20
[10] Spirin V and Mirny L A 2003 *Proc. Natl. Acad. Sci. USA.* **100** 12123-8
[11] Jeong H, Mason S P, Barabasi A L, Oltvai Z N 2001 *Nature* **411** 41-2
[12] Oyama T, Kitano K, Satou K, and Ito T 2002 *Bioinformatics* **18** 705–14
[13] Pereira-Leal J B, Enright A J and Ouzounis C A 2004 *Proteins.* **54** 49–57
[14] Saito R, Suzuki H and Hayashizaki Y 2003 *Bioinformatics* **19** 756–63
[15] Barabasi A L, Gulbahce N and Loscalzo J 2011 *Nat. Rev. Genet.* **12** 56-68
[16] Cantini L, Caselle M, Fortunato S and Medico E 2015 *Sci Rep.* **5** 17386
[17] Li W, Dai C, Liu C C and Zhou X J 2012 *J Comput Biol.* **19** 710–73
[18] Schonbach C 2009 Molecular Biology of Protein-Protein Interactions for Computer Scientists *Biological Data Mining in Protein Interaction Networks* (Japan: Kyoto university) Chapter 1 pp 1-13
[19] Ito T, Chiba T, Hattori M, Ozawa R, Sakaki Y and Yoshida M 2001 *Proc. Natl. Acad. Sci. USA.* **98** 4569–74
[20] Li X, Foo C and Ng S 2007 *Comput Syst Bioinformatics Conf.* **6** 157–68
[21] Watts DJ and Strogatz SH 1998 *Nature* **393** 440–2
[22] Bader G and Hogue C 2003 *BMC Bioinformatics* **4** 2

[23] Wuchty S and Almaas E 2005 *BMC Evol. Biol.* **5** 24.
[24] Pereira-Leal J B, Enright A J and Ouzounis C A 2004 *Proteins Struct. Funct. Bioinf.* **54** 49–57
[25] Brohee S and van Helden J 2006 *BMC Bioinf.* **7** 488
[26] Enright A J, Van Dongen S, Ouzounis C A 2002 *Nucleic Acids Res* **30** 1575-84
[27] KeshavaT S *et al.* 2009 *Nucleic Acids Res.* **37** 767–72
[28] Van Dongen S 2000 Graph Clustering by Flow Simulation (Thesis) (Netherlands : University of Utrecht)
[29] Yoon J, Blumer A and Lee K. 2006 *Bioinformatics* **22** 3106–8
[30] Liu G M, Chua H N and Wong L 2009 *Bioinformatics.* **25** 1891–7
[31] Frey B J and Dueck D 2007 *Science.* **315** 972-6
[32] Tong A H *et al.* 2004 *Science.* **303** 808-13
[33] Paccanaro A, Casbon J A and Saqi M A 2006 *Nucleic Acids Res.* **34** 1571-80

**Acknowledgment**
The authors express their gratitude to Bioinformatics research group, Computer Science department for registration of this paper and search related to this study.

IOP Conf. Series: Journal of Physics: Conf. Series **835** (2017) 012002 doi:10.1088/1742-6596/835/1/012002

# Marker traits association of agronomical traits correlated with stagnant flooding tolerance in rice

**T Sitaresmi[1,2*], D W Utami[3], W B Suwarno[4], S W Ardie[4], U Susanto[2], and H Aswidinnoor[4]**

[1]Plant Breeding and Biotechnology, Graduate School of IPB, Bogor, 16680,Indonesia,
[2]Indonesian Center for Rice Research, Subang, 41256,Indonesia
[3]Indonesian Centre for Agricultural Biotechnology and Genetic Resources Research and Development, Bogor, 16111, Indonesia
[4]Departement of Agronomy and Horticulture, IPB, Bogor, 16680, Indonesia

E-mail: sitares_trias@yahoo.com

**Abstract**. In deep-water areas, the water depth increases gradually throughout the year and maintains up to more than 50 cm of deep of water for long period. In these situations, elongation ability is necessary to allow the plants to keep up with rising floodwater.The elongation of internode during submergence is regulated by environmental and hormonal factors. The objective of this study was aimed to identify the SNP markers on 384 SNPs linked with agronomical and morphological traits related to stagnant flooding tolerance. The research were conducted at Indonesian Center for Rice Research and Indonesian Centre for Agricultural Biotechnology and Genetic Resources Research and Development. The phenotypical data was collected from F2 from bi-parental crossing of IR 42 and IRRI 119. IR 42 was sensitive parent, and IRRI 119 was tolerant. DNA extraction for rice was using a modified version of Murray and Thompson method using cetyl tri-methyl-ammonium bromide (CTAB). The genotyping was carried out using 384 SNPs Golden Gate Illumina assay. Association analysis between SNP markers and phenotypical data was performed using General Linear Model in Tassel versus 5.0 software program. Based on GLM analysis, the significant marker for plant height with P value < 0.05 are TBGI275345, TBGI275367, and TBGI424383. The significant marker for number of tiller are TBGI000722, TBGI258600, TBGI270843, TBGI271066, TBGI271076, TBGI272122, TBGI272241, and TBGI327790. Two of them, TBGI424383 and TBGI271066 were expected associated with family of protein kinase which play role in plant stress signalling.

## 1. Introduction

Rice has a unique ability to elongate its internode with increasing water depth. During submergence, rice promotes growth of shoots toward the air-water surface which facilitates gas exchange with atmosphere [1]-[2]-[3]-[4]. Stagnant flooding at depths greater than 25 cm adversely affects the growth of rice in medium-deep areas (< 100 cm), even though the plants are not fully submerged. In particular, it reduces tillering and increases lodging [5]-[6].

The grain yield-related rice traits are controlled by multiple genes which are significantly influenced by environment [7]-[8]. If the amount and distribution of favourable genes are unknown, it is difficult

* Corresponding author

IOP Conf. Series: Journal of Physics: Conf. Series **835** (2017) 012002 doi:10.1088/1742-6596/835/1/012002

to apply the rice germplasm in conventional breeding. There are interaction between genes/QTLs underlying yield-related traits and particular ecological environments. For instance, the efficiency of certain favourable genes can be enhanced by proper environments, but countered by unsuitable environments [9]-[10]. Thus, it is important and prerequisite to study favourable trait-related genes in specific ecological environment for utilizing the rice germplasms.

Several studies have been conducted on the inheritance of early elongation ability [11]-[12]-[13]. The results differed from cross to cross, apparently because rice is traitized by a variety of phylogenetic backgrounds [14]. Two quantitave trait loci (QTLs) that control the early elongation ability on chromosomes 3 and 12 in F2 population. Another study performed a diallel analysis of early elongation ability showed that some additional recessive allele(s) might confer the earlier elongation [15].

Single nucleotide polymorphisms (SNPs) can be converted into genetic markers that are scored in mapping populations using various high-throughput SNP-typing technologies [16]-[17]-[18]. High-throughput SNP discovery [19]-[20]-[21]-[22]-[23]-[24] and genotyping technologies have simplified the generation of genetic maps and the analysis of recombinants [25]. Dense maps in economically important crops will be invaluable for marker-assisted selection programs [26], analyzing linkage disequilibrium [27], detection of intra-species cis-regulatory variation [28], and other quantitative genetic studies [29].

The elongation of internode, as well as the degree of elongation during submergence is regulated by environmental and hormonal factors *viz*ethylene, GA, and ABA [3]. Further, reference [30] reported that this traits is controlled by two major SK genes driven by the rice actin promoter. There are few reports available that specify the pattern of internode elongation under deep-water. However, very scanty report are available on response of physiological (leaf area, photosynthesis, and non-structural carbohydrates) and morphological (shoot length, tillers, culm thickness, spikelet fertility, shoot biomass, and grain yield) traits under deep-water rice [31]-[32]-[33].

The objective of the study is to select markers associated with agronomical traits of rice related to stagnant flooding tolerance using SNPs marker

## 2. Methodology

Analysis was conducted in Indonesian Centre for Rice Research (for DNA extraction) and Molecular Biology Laboratory of Indonesian Centre for Agricultural Biotechnology and Genetic Resource (for genotyping assay) on February to June 2016. Genetic materials used are population of P1 (IR 42), P2 (IRRI 119), and F2.

As markers, we used 384 SNPs linked with morphological and agronomical traits. All markers used were validated by Rice Genome Division of Indonesian Center for Agricultural Biotechnology and Genetic Resource [34]. Phenotypic data associated with the genotypic data were plant height, number of tiller, weight of 100 grain, panicle length, and grain yield. Panicle length, weight of 100 grain, and grain yield were selected based on plant height and number of tiller is the character that can be used as selection criteria for stagnant flooding tolerance. Plant height and number of tiller are characters related to stagnant flooding tolerance based on [31]-[35]-[36]-[37]. Weight of 100 grains, panicle length, and grain yield were determined as selection criterion for stagnant flooding tolerance. Background selection method was used in this experiment. There are two steps for genotyping, namely DNA extraction and Genotyping Assay.

### *2.1 DNA extraction*

DNA extraction for rice was using a modified version of a method developed by Murray and Thompson [38]. It is a simple effective method utilizing cetyl tri-methyl-ammonium bromide (CTAB).

Materials used were 2-3 inches leaves of 3-week old plant (one leaf per sample), scissors, sterile 2 ml micro tubes, steel balls, liquid nitrogen, micropipettes, tips and tubes, water bath, centrifuge with rotor, -20 $^{0}$C freezer, 800 μl 2x CTAB buffer warmed to 65 $^{0}$C, 0.8 ml chloroform-isoamyl alcohol

(24:1), 600 µl isopropanol, 70 % ethanol, 200 µl TE buffer, 2 µl RNAse (10 mg/ml), 20 µl (1/10 volume solution) sodium acetate, 400 µl (2 volumes) absolute ethanol.

DNA of genotype was bulked based on phenotypic clustering. Technique of sampling which will be used is bulked segregant analysis. Bulkedsegregant analysis (BSA) has been proposed as anefficient strategy for identifying DNA markers linkedto the genes or genomic regions of interest [39].BSA has been successfully used in rice for identifyinglarge effect QTLs linked to rice yield under severedrought stress [40]-[41].

Dilution will be done to obtain the required concentration of DNA in Genotyping Assay. DNA diluted by ddH2O up to DNA concentration is suitable with its concentration for Genotyping Assay. Dilution formula is V1 x V2 x M1 = M2, with V1 is DNA volume before dilution, M1 is DNA concentration before dilution, V2 is final volume dilution, and M2 is concentration DNA for Genotyping Assay, that is 25 ng/mL.

### *2.2 GoldenGate Genotyping Assay*

GoldenGate Genotyping Assay consisted of two process: pre-amplification and post-amplification. Pre-amplification consisted of DNA activation, precipitation, resuspension, ASE (*Allele Specific Extension),* MEL (*Master mixed Enzymatic Extension and Ligation*), PCR plate preparation, and Inoc PCR. DNA activation process of each samples used random biotinylated with added MS1 reagent. Thus precipitation was done by adding PS1 reagent and 2-propanol and was continued resuspension by adding RS1 reagent. ASE process was by adding biotinylated gDNA and query oligos, hybridization reagent, and paramagnetic particle. Then it was continued with washing the product of non-specific hybridization and oligo exceed by adding AM1 and UB1 reagent. Then process of enzymatic extension and ligase was done by adding MEL master mix to DNA samples.

Preparation of PCR plate for amplification was by adding Titanium Taq DNA polymerase, and then sample was removed to PCR plate. Extension and ligase process were adding PCR (Ioc PCR) using 3 universal primer (reagent MMP), i.e. two primers with fluorescence labeled (Primer 1 and Primer 2), and 1 Primer biotinylated (Primer 3), where Primer 3 was tag the PCR product and to elute DNA strand which contained fluorescence signal. Then eluted sample was transferred from ASE plate to PCR plate.

Post amplification process consisted of PCR, binding PCR, INT plate for BeadChip, hibydrization, washing BeadChip, and visualization of BeadChip on IScan system. PCR process was amplification of PCRA plate with fluorescence label, then was continued with binding PCR. Binding PCR was done by adding MPB reagent to PCR plate and then the suspension was transferred to plate filter. DNA strand was biotinylated with paramagnetic particle so that produced immobile double strand PCR product. In the INT plate for BeadChip step, PCR product with SS fluorescence was washed with UB1, MH1 and NaOH 0.1N and then was dissolved in intermediate (INT) plate. BeadChip was hybridize using Hyb Chamber for one night in ilumina hybridization oven at 60°C to 45°C. BeadChip was transferred and washed using PB1 (2x) and XC4 (1x). Visualization of BeadChip on IScan Reader system use laser to excited Fluor form single-base extension product in bead. Light emission from the Fluor was recorded as picture with high resolution. The visualization product was analyzed to determining SNP genotype using Illumina's BeadStudio Gene Expression Module [42].

### *2.3 Data analysis*

Phenotypical data used were plant height and number of productive tiller of P1, P2, and F2 population. A number of 5% of minimum and maximum value of plant height and number of tiller from F2 population was analysis using t test to confirm the phenotypic variability. Association analysis between SNP marker and phenotypical data was tested using General Linear Model in Tassel 5 software [43]. Values of the Q matrix obtained in Structure were presented as covariates. The P value determines whether a marker was associated with phenotype and $R^2$ for marker evaluates the magnitude of QTL effect to phenotype.

## 3. Result And Discussion

DNA was bulked based on cluster analysis in phenotypic data of plant height and number of tiller. Plant height is the main character to avoid plant fromstagnant flooding stress or escape strategy [44]. Length of stem was identified determine the stagnant flooding tolerance. Stem length have high correlation with plant height. Meanwhile, number of tiller is character which responsive to stagnant flooding stress [45]-[35]-[37]. Both of these characters can be observed at the end of the vegetative stage. Therefore, cluster analysis was performed on the characters. Others characters observed in the generative stage or physiological maturity stage were panicle length and weight of 100 grains.

We used the T-test to determine whether there is phenotypic variability observed in F2 population (Table 1). T test was performed on plant height and number of tillers. T test showed that that the average of plant height and number of productive tiller across 10% of minimum values (left extremes) were significantly different from those across 10% of maximum values (right extremes) (Figure 1 and 2).

**Table 1.** T test of minimum and maximum value of plant height and number of tiller from F2 population

| Variable | N | Mean | StDev | SE Mean | P-Value |
|---|---|---|---|---|---|
| Plant height | | | | | |
| PH_min | 10 | 123.1 | 1.37 | 0.43 | 0.000 |
| PH_max | 10 | 161.0 | 4.55 | 1.40 | |
| Number of tiller | | | | | |
| PT_min | 10 | 2.7 | 1.06 | 0.33 | 0.000 |
| PT_max | 10 | 18.3 | 1.06 | 0.33 | |

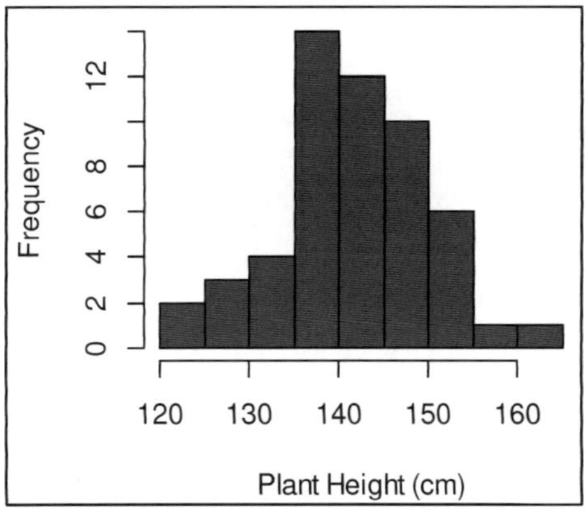


**Figure 1.** Distribution of plant height and number of tillers of rice group

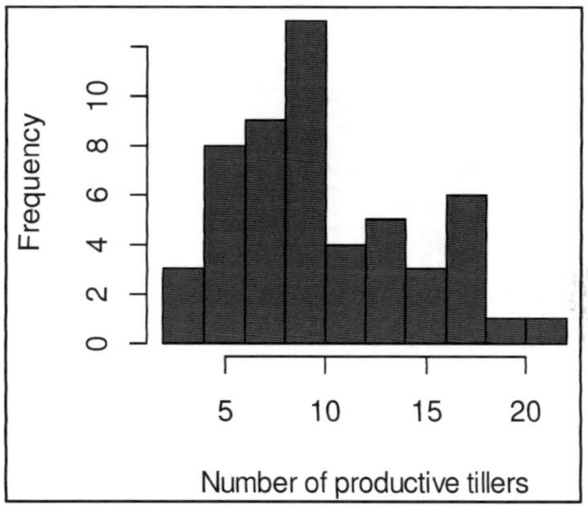


**Figure 2.** Distribution of number of tillers of rice group

Marker which showed polymorphic on 53 rice sample were 82 of 384 SNPs marker used. Association mapping analysis using GLM showedthat there were five SNPs associated with plant height with probability value <0.05 and $R^2$was ranged between 12-13%. Likewise, mapping association between the markers and the number of productive tillers showed 10 marker have probability value < 0.05 and $R^2$was ranged between6-15% (Table 5.2). Association mapping for panicle length showed 9 SNPs markers were significant with probability < 0.01 and $R^2$ was ranged 18-

28 %. $R^2$ showed magnitude of QTLs effect controlling the phenotype. Although it was detected with less statistical power, MAF<0.05 but > 0.02 is still appropriate to help in identification of smaller set of candidates genes using bi-parental population [46].

Based on General Linear Model (GLM) association analysis, significant genomic regions for plant height and number of tiller were identified based on P value < 0.05. The significant markers for plant height were TBGI275345, TBGI275367, dan TBGI424383. TBGI275367 was expected on same position with LOC Os06g09630 which probably related with 3-oxoacyl-synthase gene. LOC Os06g09630 were functionally annotated as chloroplast precursor putatively expressed genes on the basis of the homology with the previous annotated genes. Chloroplasts have been observed to be involved in light perception and interact with photoperiodic development acting as major light harvesting organs [47]. TBGI424383 is located in chromosome 11 and was mapped near from probably gene controlling serine protease. Serine protease have been shown to be involved in diverse processes regulating plant development and defence response to abiotic and biotic stress [48]-[49]-[50]-[51].

The significant SNPs for number of tiller were TBGI000722, TBGI258600, TBGI270843, TBGI271066, TBGI271076, TBGI272122, TBGI272241, and TBGI327790. The SNPs marker TBGI271066 is associated with OsWAK61 gene. The Wall-Associated Kinase (WAK) is a subfamily of receptor-like kinase (RLK), with some members identified as associated to cell wall, suggesting these genes are strong candidates to act as sensors directly linking the extracellular environment to the cytoplasm and triggering intracellular signals. TBGI327790 was associated with Glucan endo-1,3-beta glucosidase. Glucan endo-1,3-beta glucosidase protein that were up regulated under stress [52]. Proteins involved in multiple pathways showed significant changes in expression in response to stress environment, including oxidative stress response proteins, beta-glucanases, protein; kinases, glycine, and photosynthesis proteins. Signal transduction is an important molecular response which plays a major role in protecting the plants by modulating the stress-responsive gene expression. These signals may be transmitted to neighbouring cells/plants by emitting volatile organic compounds [53]. TBGI270843 was allegedly in same position with LOC_Os06g04810.1which is associated with leucine rich repeat protein gene.

**Table 2.** SNP marker significantly associated with plant height and no of tiller under stagnant flooding stress

| Trait/marker | Chr | Position | P value | $R^2$ | Nearby gene candidate |
|---|---|---|---|---|---|
| Plant Height | | | | | |
| TBGI275345 | 6 | 4869776 | 0.036 | 0.129 | - |
| TBGI275367 | 6 | 4897700 | 0.038 | 0.130 | 3-oxoacyl-synthase, chloroplast precursor |
| TBGI424383 | 11 | 6199852 | 0.037 | 0.128 | Serine protease-like proteins |
| No. of tiller | | | | | |
| TBGI000722 | 1 | 421982 | 0.031 | 0.113 | Expressed protein |
| TBGI258600 | 5 | 25247484 | 0.039 | 0.106 | Glucose 1 dehydrogenase (GlcDH) |
| TBGI270843 | 6 | 2096700 | 0.038 | 0.107 | Leucine rich repeat protein, putative, expressed |
| TBGI271066 | 6 | 2229688 | 0.034 | 0.110 | OsWAK61 - OsWAK receptor-like protein kinase |
| TBGI271076 | 6 | 2234192 | 0.018 | 0.129 | OsEF3 |
| TBGI272122 | 6 | 2829380 | 0.044 | 0.067 | - |
| TBGI272241 | 6 | 2877413 | 0.01 | 0.152 | - |
| TBGI327790 | 7 | 21148555 | 0.034 | 0.110 | Glucan endo-1,3-beta glucosidase precursor |

The most significant marker for the characters was described with QQ plot (Figure 3). QQ plot is a visualization of Table 5.1. Significant marker was showed by highest Log 10(P-value). The SNP marker associated with number of tiller, TBGI272241, which located in chromosome 6 at 2877413bp. TBGI2241 is not yet uncharacterized, so that its gene has not been identified yet. This provides an opportunity to identify a new gene associated with number of tiller in relationship with stress water tolerance.

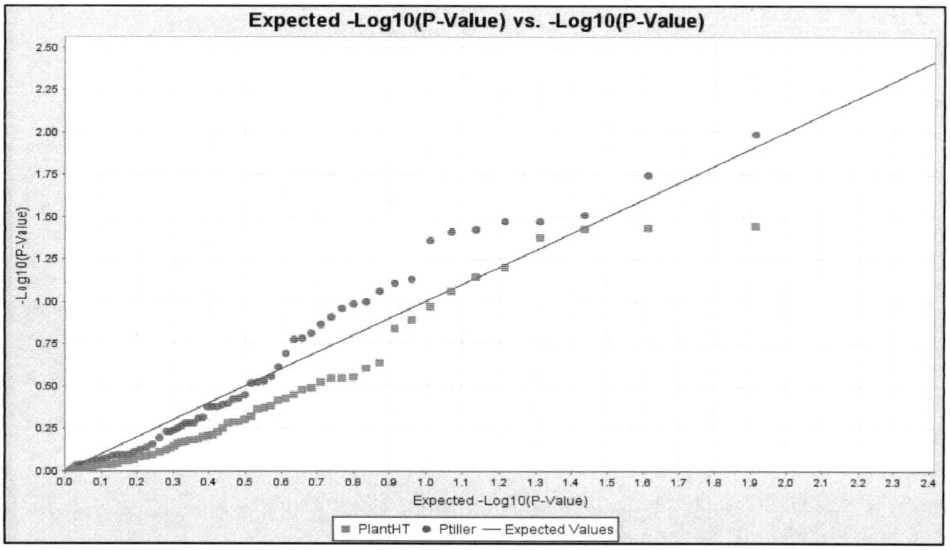


**Figure 3.** Quartile-quartile plot (qq plot) determines marker-traits association compare to the expected result

The ability of multiple linear regression model to explain proportion of phenotypic variance could be used to identify simple assay involving a small number of SNPs which are designed for selecting lines with favourable alleles for target traits [46]. A linear model with two candidate marker explained 13.3 % of phenotypic variance for number of tiller. Two marker involving the regression model for number of tiller were TBGI270843 and TBGI272241 (Table 3 and 4). Meanwhile for plant height, regression model could explained 10 % of variance, which were contributed by SNPs marker TBGI258567 and TBGI275345.

**Table 3.** Anova of multiple linear regression of tiller number, plant height based on selected SNP

| Source | DF | Mean Square | Pr(> F) | Adj $R^2$ |
|---|---|---|---|---|
| Number of tillers | | | | |
| Model | 2 | 88.211 | 0.012 | 0.133 |
| Error | 48 | 18.227 | | |
| Total | 50 | | | |
| Plant height | | | | |
| Model | 2 | 243.629 | 0.027 | 0.102 |
| Error | 49 | 62.525 | | |
| Total | 51 | | | |

**Table 4.** Parameter Estimates

| Variable | Estimate | Std. Error | Pr(>\|t\|) |
|---|---|---|---|
| Tiller number | | | |
| Intercept | 9.58 | 0.72 | 0.000 |
| TBGI270843 | -5.37 | 1.85 | 0.006 |
| TBGI272241 | 3.92 | 1.43 | 0.008 |
| Plant height | | | |
| Intercept | 145.47 | 2.6 | 0.000 |
| TBGI258567 | 5.01 | 2.69 | 0.068 |
| TBGI275345 | -4.82 | 2.4 | 0.050 |

## 4. Conclussion

Based on General Linear Model (GLM) association analysis, the significant marker for plant height with P value < 0.05 were TBGI275345, TBGI275367, and TBGI424383. The significant marker for number of tiller were TBGI000722, TBGI258600, TBGI270843, TBGI271066, TBGI271076, TBGI272122, TBGI272241, and TBGI327790. TBGI424383 and TBGI271066 probably associated with family of protein kinase which play role in plant stress signaling.

## References

[1] Kende H K E and Cho H T 1988 *Plant Physiol.***118** 1105-10
[2] Sauter M 2000 *Natur wissen schaften.* **87** 289-03.
[3] Vriezen W H, Zhou Z and Straeten D 2003 *Ann Bot.* **91** 263-70.
[4] Jackson M B and Ram P C 2003 *Ann Botani.* **91** 227-41
[5] Singh S and Singh R K 2000 *Rainfed Rice: A Source Book of Best Practices and Strategies in Eastern India* (Los Banos, Phillipines. International Rice Research Institute)
[6] Tuong T P, Kam S P, Wade L, Pandey S, Bouman B A M and Hardy B 2000 *Traitizing and Understanding Rainfed Environments* (Los Banos, Philippines. International Rice Research Institute)
[7] Huang R Y, Jiang L R, Zheng J S, Wang T S, Wang H C, Huang Y M and Hong Z L 2013 *Trends Plant Sci.* **8** 218-26
[8] Ikeda M, Miura K, Aya K, Kitano H and Matsuoka M 2013 *Curr. Opin Plant Biol.* **16** 213-20
[9] Liu G F, Yang J and Zhu J 2006 *Acta Genet Sin.* **33** 607-16
[10] Sreedhar S, Reddy T D and Ramesha M S 2011 *Int J Plant Breeding Genet.* **125** 1303-12
[11] Tripathi R S and Rao M J B 1985 *Euphytica.* **34** 875-81.
[12] Suge H 1987 *Jpn. J. Genet* **62** 69-80
[13] Eiguchi M, Hirano H Y, Sano Y and Morishima H 1993 *Jpn. J. Breed.* **43** 135–39.
[14] Glaszamann J C 1987 *Theor. Appl Genet.* **74** 21-30.
[15] Nemoto K, Ukai Y, Tang D Q, Kasai Y and Morita M 2004 *Theor. Appl. Genet* **109** 42-7
[16] Gabriel S and Ziaugra L 2004 *Curr. Protoc. Hum Genet* **60** 2.12.1- 2.12.18
[17] Gunderson K L, Steemers F J, Lee G, Mendoza L G and Chee M S 2005 *Nat. Genet.* **37** 549–54
[18] Hui L, DelMonte T and Ranade K 2008 *Curr. Protoc. Hum. Genet.* **56** 2.10.1-2.10.8.
[19] Marth G T, Korf I, Yandell M D, Yeh R T and Gu Z 1999 *Nat. Genet.* **23** 452–6
[20] Weckx S, Del-Favero J, Rademakers R, Claes L and Cruts M 2005 *Genome Res.* **15** 436–42
[21] Zhang J, Wheeler D A, Yakub I, Wei S and Sood R 2005 *PLoSComput. Biol.*1 e53
[22] Barbazuk W B, Emrich S J, Chen H D, Li L and Schnable P S 2007 *Plant J.* **51** 910–8
[23] Li H, Ruan J and Durbin R 2008 *Genome Res.* **18** 1851–8.
[24] Li R, Li Y, Fang X, Yang H and Wang J 2008 *Genome Res.* **19** 1124–32
[25] Shifman S, Bell J T, Copley R R, Taylor M S and Williams R W 2006 *PLoS Biol.* 4 e395

[26] Prigge V, Melchinger A E, Dhillon B S and Frisch M 2009 Theor. *Appl. Genet.* **119** 23–32
[27] Kruglyak L 2008 *Nat. Rev. Genet.* **9** 314–8
[28] Stupar R M and Springer N M 2006 *Genetics* **173** 2199–210
[29] Cookson W, Liang L, Abecasis G, Moffatt M and Lathrop M 2009 *Nat. Rev. Genet.* **10** 184–94
[30] Hattori Y, Nagai K, Furukawa S, Song X J, Kawano R, Sakakibara H, Wu J Z, Matsumoto T, Yoshimura A, Kitano H, Matsuoka M, Mori H and Ashikari M 2009 *Nature* **460** 1026–30
[31] Amante M M 1986 Evaluation of rice breeding lines under medium-deep (25-50 cm) water conditions (Thesis) (Gregorio: Aranante University)
[32] Sakagami J I, Joho Y and Ito O 2009 *Ann. Bot.* **103** 171-80
[33] Vergara G V, Nugraha Y, Esguerra M Q, Mackill D J and Ismail AM 2014 *Aob PLANTS* **6** plu055
[34] BB Biogen 2015 Pengembangan TOOL Kits Sistem Deteksi Karakter Penting Tanaman Padi Unggul. Tim Genom Padi (Laporan Hasil Penelitian) (Bogor : Balai Besar Penelitian Bioteknologi dan Sumberdaya Genetik Pertanian)
[35] Singh S, Mackill D J and Ismail A M 2011 *Field Crops Res.* **121** 311-23
[36] Collard B C Y, Septiningsih E M, Das S R, Carandang J J, Pamplona A M, Sanchez D L, Kato Y, Ye G, Reddy J N, Singh U S, Iftekharuddaula K M, Venuprasad R, Vera-Cruz, Mackill D J and Ismail A M 2013 *SABRAO J. Breed. Genet.* **45** 42-56
[37] Kato Y, Collard B C Y, Septiningsih E M, and Ismail A M 2014 *Aob PLANT* .**6** plu058
[38] Murray M G and Thompson W F 1980 *Nucleic Acids Res.* **8** 19
[39] Brauer M J, Christianson C M, Pai D A and Dunham M J 2006 *Genetics* **173** 1813–16
[40] Quarrie S A, Lazic-Jancic V, Kovacevic D, Steed A and Pekic S 1999 *J. Exp. Bot.* **50** 1299–306
[41] Venuprasad R, Dalid C O, Del Valle M, Zhao D, Espiritu M, Sta Cruz M T, Amante M, Kumar A and Atlin G N 2009 *Theor. Appl Genet.* **120**177–90
[42] Illumina 2008 Retrieved from http://support.illlumina.com/documents.
[43] Bradbury P J, Zhang Z, Kroon D E and Bucker E S 2007 *Bioinformatics.* **23** 2633-2635
[44] Hattori Y, Nagai K and Ashikari M 2011 *Curr. Opin. Plant Biol.* **14** 100-5.
[45] Ismail AM, Johnson DE, Ella ES, Vergara GV, and Baltazar AM 2012 *Aob PLANTS* **2012** pls019
[46] Suwarno W B, Pixley K V, Palacios-Rojas N, Kaeppler S M and Babu R 2015 *Theor Appl. Genet.***128** 851-64
[47] Lepistö A, Kangasjärvi S, Luomala E M, Brader G, Sipari N, Keränen M, Keinänen M and Rintamäki E 2009 *Plant Physiol.***149** 1261–76
[48] Nakagami H, Pitzschke A and Hirt H 2005 *Trends Plant Sci.* **10** 339-46.
[49] Schaller A 2004 *Planta* **220** 183-197
[50] Adam Z, Adamska I, Nakabayashi K, Osterstzer O, Haussuhl K, Manuell A, Zheng B, Vallon O, Rodermel S R, Shinozaki K and Clarke A K 2001 *Plant Physiol.* **125** 1912-8
[51] Palma J M, Sandalio L M, Corpas F J, Romero-puertas M C, McCarthy I, Rio L A 2002 *Plant Physiol. Biochem.* **40** 521-30
[52] Chen X, Dong Y, Yu C, Fang X, Deng Z, Yan C and Chen J 2016 *PloS One* **11** 1-16
[53] Bruinsma J J, Schneider D L, Davis D E and Kornfeld K 2008 *Genetics.***179** 811-28.

**Acknowledgments**
Authors gratefully acknowledge to SEARCA Scholarship for financial support to conduct this study, Indonesian Center for Rice Research (ICRR), and Indonesian Centre for Agricultural Biotechnology and Genetic Resources Research and Development (ICABIOGRAD) for providing laboratory facilities.

# Prediction of maize phenotype based on whole-genome single nucleotide polymorphisms using deep belief networks

**H Rachmatia*, W A Kusuma, and L S Hasibuan**

Department of Computer Science, Bogor Agricultural University, Indonesia

E-mail: hikmat_rac12t@apps.ipb.ac.id

**Abstract**. Selection in plant breeding could be more effective and more efficient if it is based on genomic data. Genomic selection (GS) is a new approach for plant-breeding selection that exploits genomic data through a mechanism called genomic prediction (GP). Most of GP models used linear methods that ignore effects of interaction among genes and effects of higher order nonlinearities. Deep belief network (DBN), one of the architectural in deep learning methods, is able to model data in high level of abstraction that involves nonlinearities effects of the data. This study implemented DBN for developing a GP model utilizing whole-genome Single Nucleotide Polymorphisms (SNPs) as data for training and testing. The case study was a set of traits in maize. The maize dataset was acquisitioned from CIMMYT's (International Maize and Wheat Improvement Center) Global Maize program. Based on Pearson correlation, DBN is outperformed than other methods, kernel Hilbert space (RKHS) regression, Bayesian LASSO (BL), best linear unbiased predictor (BLUP), in case allegedly non-additive traits.DBN achieves correlation of 0.579 within -1 to 1 range.

## 1. Introduction

Traditional genetic improvement of plant breeding, using information on phenotypes and pedigrees to predict trait values, has been very successful. However, trait values should be able to predict more accurately by using information on variation in DNA sequence between plants. Marker-assisted selection (MAS) has been used in plant breeding improvement program since the 1990s, after promising research results for tagging genes or mapping quantitative trait loci (QTL). Currently, MAS has failed to significantly improve polygenic traits. While MAS has been effective for the manipulation of large effect alleles with known association to a marker, it has been at a deadlock when many alleles of small effect separate and no substantial, reliable effects can be identified.

The introduction of Genomic Selection (GS) has paved the way to overcome these limitations. GS has shifted paradigm in MAS that seeking to identify individual loci significantly associated with trait into paradigm that uses all marker data as predictors of performance and consequently delivers more accurate predictions. Selection can be based on the output of genomic prediction (GP) models. GP uses a training population of individuals that have been both genotyped and phenotyped to develop a model that takes genotypic data from a candidate population of untested individuals and produces estimated trait values. This approach potentially leading to more rapid and lower cost gains from plant breeding [1].

* Corresponding author

Building an appropriated GP models poses several statistical and computational challanges, such as how models can cope with the curse of dimensionality, colinearity between markers, or the complexity of quantitative traits that involves non-additive effects. Linear model in parametric methods (i.e Bayessian LASSO (BL) [2], and best linear unbiased predictors (BLUP) [3]), that are frequently used in developing GP model typically ignore gene by gene interactions, as well as higher order non-linearities [4]. To meet this challange, and to take possible non-linearities into account in prediction, there has been a growing interest in the use of semiparametric and nonparametric methods. In this context, machine learning (as one of nonparametric methods) have been considered to be promising predictive machineries.

Deep learning is a branch of machine learning, currently has been being the most popular topic in area of machine learning research in a last decade. Deep learning is a set of algorithm that attempt to model high level abstractions in data by using a deep graph with multiple linear and non-linear tranformation. This method succeeded to solve perceptual problems such as image recognition [5] and speech recognition [6]. Deep learning has been characterized as a buzzword, because it is just rebranding of artificial neural network.

The study about GP that involve deep learning approach have explored before. Reference [7] constructed deep learning architecture with multi-layer restricted Boltzmann machines (RBM) deep network to disease diagnosis of schizoprenia with single nucleotide polymorpishms (SNPs) molecular marker data as predictors. This method can obtain the average accuray is over 93% and outperformed than other method such as k-nearest neighbours and support vector machines (SVMs). However, that proposed method slightly down in the performance compared with deep belief networks (DBN), which is another deep learning architecture [8].

In this paper, we propose a 4-layer DBN which composed by three RBM to develop GP models. We modified the DBN such that it addresses regression problem instead of classification problem that is usually DBN used to. We used maize dataset from CIMMYT's (International Maize and Wheat Improvement Center) Global Maize program. Reference [9] used the same data to develop GP models with parametric and semiparametric methods. Parametric method consist of BL and BLUP, and then semiparametric method consist of repoducing kernel Hilbert spaces (RKHS) method [10]. To measure the performance of DBN GP models, we compared DBN and these three methods based on Pearson's correlation in the end.

## 2. Methodology

### *2.1 SNPs Data and Preprocessing*

The maize data set is from the Dought Tolerance for Africa project of CIMMYT's Global Maize Program. The original data set included 300 tropical lines genotyped with 1148 SNPs (referred to as markers). For each marker, the alleles with lowest frequeny was coded as one.

Trait analyzed for this study were grain yield (GY), female flowering (FFL) (or days to silking), male flowering (MFL) (or days to anthesis), and the anthesis-silking interval (ASI), each evaluated under severe drought stress (SS) and well-watered (WW) conditions. Hereinafter we refer to these data sets as maize-grain yield (M-GY) and maize-flowering (M-F), repectively. In all environments, the response variable was standardized to a sample variance equal to one.

The number of lines in the M-F data set was 284, whereas 264 lines were available in M-GY. The average minor allele frquency in these data sets was 0.20. Marker with allele frequency less than 0.05 or greater than 0.95 were removed. Missing genotypes were imputed using samples from the marginal distribution of marker genotypes, that is, $x_{ij} \sim Bernoulli(\hat{p}_j)$, where $\hat{p}_j$ is the estimated allele frequency computed from nonmissing genotypes [9]. After edition, the numers of markers available for analysis were 1148 and 1135 in M-F and M-GY, respectively.

### *2.2 Deep Belief Networks*

Deep belief network (DBN) was introduced by Hinton [8] as a probabilistic generative model which contain multi layers of hidden variable, and these layers detect the higher order of strong corelation between activity of hidden feature in its above layer. State of the art from DBN is applying greedy layer-wise unsupervised learning algorithm to pre-training the initial weights of the networks, and then with global supervised learning for fine-tuning the networks. We used RBM as pre-training algorithm for the DBN.

An RBM is bipartite graph in which visible units that represents observations are connected to binary, stochastic hidden units using undirected weighted connections. They are restricted in the sense that there are no visible-visible or hidden-hidden connections. RBM has an efficient training procedure which makes them suitable as building blocks for DBN.

The input of the first layer in the DBN is the markers data, i.e, the information is coded as -1, 0, or 1. Denote the length of each sample as $N_v$. In this paper, $N_v = 1148$ for M-F data set and $N_v = 1135$for M-GY data set. In order to obtain more accurate feature selection than generally used softmax RBM for the multi-states sequences, firstly, the marker sequences are changed into binary sequences [7].

We developed a 4-layer DBN which was contained of a visible layer and 3-layer RBM. SNPs data are available from data set that are used as the raw input for the visible layer. Next, each RBM layer can find a feature representation of the current input data, and finally the last layer of the model provides abstract features of the raw features for the regression. The overall approach to building and training a DBN to predict quantitative trait value from markers data is shown in Figure 1.

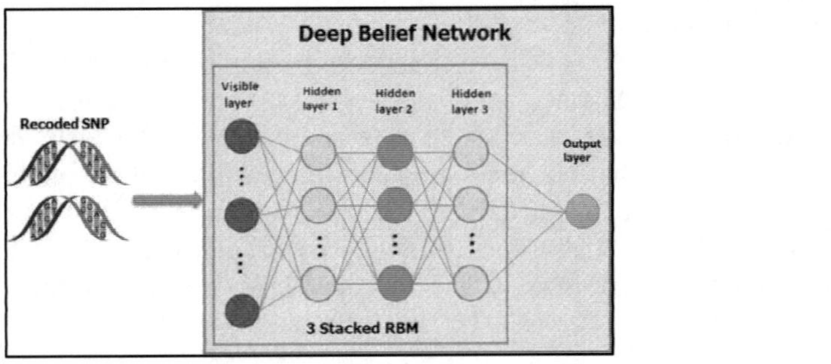


**Figure 1.** The DBN architecture for genomic prediction

In an RBM, thereare are two layers; one is the layer of visible units, and the other is the layer of hidden units. There are connections between the layer but no connection between units within each layer. In order to get beter performance, we do not choose the general used RBM with softmax and multinomial units, instead we use the Bernoulli RBM as in [7] and define the energy function as follows:

$$E_\theta(v,h) = -v^T Wh - \alpha^T v - \beta^T h \quad (1)$$

where the visible units $v = (v^{(-1)}, v^{(0)}, v^{(1)})$, and $v^{(i)} = (v_1^{(i)}, v_2^{(i)}, \dots, v_{N_v}^{(i)})^T$ $(i = -1, 0, 1)$ is a $N_v$-dim vector with each $v_j^{(i)} \in \{0,1\}$. Hidden units denoted as $h = (h_1, h_2, \dots, h_{N_h})^T$with each $h_k \in \{0,1\}$. Bias of the visible units is $\alpha = \left(a_1, a_2, \dots, a_{3*N_v}\right)^T$, bias of the hidden units is $\beta = \left(b_1, b_2, \dots, b_{N_h}\right)^T$, and $W = \{w_{jk}\}_{4*N_v x N_h}$ with each $w_{jk}$ is the connection of $v_j$ and $h_k$. Here $\theta$ is the parameters of the RBM, especially refering to the connection weight matrix $W$ between the visible units $v$ and hidden units $h$, the bias $\alpha$ of $v$, and the bias $\beta$ of $h$. For every line sample, denote the SNPs sequence as $S$, i.e, $S$ is a $N_v$-dim vector and each $S_j \in \{-1, 0, 1\}$. Define $v_j^{(i)}$ $(i = -1, 0, 1;\ j =$

$1, 2, \dots, N_v)$ as follows:

$$v_j^{(i)} = \begin{cases} 1, & S_j = i \\ 0, & S_j \neq i \end{cases} \tag{2}$$

Thus, we have the joint probability distribution of $(v, h)$ as follows:

$$P_\theta(v, h) = \frac{1}{Z_\theta} exp(-E_\theta(v, h)) \tag{3}$$

where $Z_\theta = \sum_{v,h} exp(-E_\theta(v, h))$ is known as the partition function or normalizing constant.

The distribution of the observed data $v$, $P_\theta(v)$, is

$$P_\theta(v) = \frac{1}{Z_\theta} \sum_h exp(-E_\theta(v, h)) \tag{4}$$

which has another name, i.e, log-likelihood function. Let $\mathcal{L}(\theta) = -\log P_\theta(v)$. Minimizing the log-likelihood function is equal to determining the parameters to fit the given training samples, i.e, to find a $\theta^*$, such that

$$\theta^* = \arg\min_\theta \mathcal{L}(\theta) \tag{5}$$

By minimizing $\mathcal{L}(\theta)$, we get the parameters of an RBM, and this can be achieved by performing stochastic steepest descent. The RBM parameters $(\theta)$ are updated by

$$\Delta\theta = \epsilon \cdot \frac{\partial \mathcal{L}(\theta)}{\partial \theta} \tag{6}$$

in which, $\epsilon$ is the learning rate.

In this study, we use the free energy function in [11] to estimate the gradient of log-likelihood. This function defined as follows:

$$F(v) = \alpha^T v - \sum_j \log(1 + \exp(W_{j\circ} \mathrm{v} + \beta_j)) \tag{7}$$

then the gradient can be written as follows:

$$\frac{\partial \mathcal{L}(\theta)}{\partial \theta} \approx \frac{\partial F(v_{data})}{\partial \theta} - \frac{\partial F(v_{model})}{\partial \theta} \tag{8}$$

Variable $v_{data}$ is an input vector and $v_{model}$ is a reconstructed input that obtained by use contrastive divergence (CD) with Gibbs sampling procedure.

The first step we implement DBN is intialize every weights of hidden layer. The initial values for this weights are uniformly sampled from a symmetric interval that depends on the activation function. We use sigmoid function as an activation, therefore the interval is $\left[-4\sqrt{6/p_{in} + p_{out}}, 4\sqrt{6/p_{in} + p_{out}}\right]$, where $p_{in}$ is the number of units in the $(i-1)$-th layer, and

$p_{out}$ is the number of units in the $i$-th layer. This initilization ensures thta, early in training, each neuron operates in a regime of its activation function where can easily be propagated both upward (activation flowing from inputs to outputs) and backward (gradient flowing from outputs to inputs) [12].

We do some modification to DBN such that the architecture can models the regression problem of genomic prediction. We do not use softmax function as activation function in output layer, instead we use linear function ($f(x) = x$). Besides, we change the basic cost function from logistic loss to mean squared error (MSE) as follows:

$$MSE = \frac{1}{n}\sum_{i=1}^{n}(\hat{y}_i - y_i)^2 \tag{9}$$

where $n$ is the number of sample in mini-batch training, $\hat{y}_i$ is the predicted value of sample $i$ by DBN, and $y_i$ is the true value of phenotype from the sample $i$.

Furthermore, in order to prevent the model from overfitting we add a regularization term to the basic cost function. The regularizer is L2 regularization and also known as weight decay. Hence, our cost function is defined as follows:

$$cost = MSE + \lambda\|\theta\|_p^p \tag{10}$$

where $\|\theta\|_p$ is

$$\|\theta\|_p = \left(\sum_{j=0}^{|\theta|}|\theta_j|^p\right)^{\frac{1}{p}} \tag{11}$$

with $p = 2$ in L2 regularization term. Beside that, we applied early stoping procedure when fine-tuning the model in training process. We used mini-batch stochastic gradient descent optimization in [13] to fine-tune the model.

We implemented all our pre-training and training algorithm concept for DBN by using Theano library for Python programming [14]. Theano is a Python library that can used to define, optimize, and evaluate mathematical expressions, especially ones with multi-dimensional arrays. Using Theano, it is possible to attain speeds rivaling hand-crafted C programming implementations for problems involving large amounts of data.

*2.3 SNPs Data and Preprocessing*

Prediction of performance of lines whose phenotypes are yet to be observed is a central problem in plant breeding. Such prediction can be used, for example, to decide which of the newly generated lines will be evaluated in field trials. Cross-validation (CV) methods were used to evaluate the ability of model to predict future outcomes. To this end, data were divided into 10 folds; this was done by using an index variable, $I_i \in \{1, \dots, 10\}$, $i = 1, \dots, n$, that randomly assigns observations to 10 disjoint folds, $F_j \in \{i : I_i = j\}, j = 1, \dots, 10$. CV predictions of the observations in the first fold, $F_1 = \{i : I_i = 1\}$, are obtained by omitting phenotypic data on all lines in the first fold. This yields CV predictions of lines in the first fold, that is, $\{\hat{y}_i : I_i = 1\}$. Repeating this task for the second, third, ..., 10th folds fields a whole set of CV predictions $\{\hat{y}_i\}_{i=1}^{n}$ that can be compared with actual observations $\{y_i\}_{i=1}^{n}$ to assess precision. We use Pearson's correlation between phenotypes and their respective predictions from the model as precision measurement. Reference [9] also used the same CV mechanism to obtain precision of the models.

## 3. Results and Discussions

The DBN that we used to develop GP models are consist of 3 RBM layer for pre-training and 1 output layer for regression, in other words the DBN has 3 hidden layers. Each hidden layer consist of 512 neurons. In order to train the RBM we use CD-$k$ with parameter $k = 1$, and number of epoch is 500.

After pre-training the DBN, we train the DBN with feed forward-backpropagation and use mini-batch stochastic gradient descent as optimization algorithm. In this study, we compare the DBN architecture based on learning rate in fine-tune step. The highest average correlation value we choose from each learning rate variation to represent the precision of the model. Here, we use learning rate 0.01 and 0.001 for training the model.

It can be seen from Figure 2, the best correlation value is obtained from learning rate 0.001 for all data set phenotype category except data set GY-SS that has the best correlation value from learning rate 0.01.

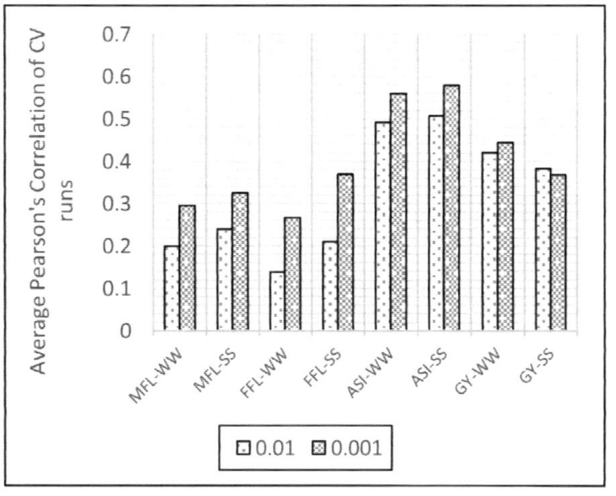


**Figure 2.** Comparison of precision for two learning rates

Therefore, we have precision of DBN for each phenotype category as follows: 0.295 (MFL-WW), 0.325 (MFL-SS), 0.267 (FFL-WW), 0.370 (FFL-SS), 0.559 (ASI-WW), 0.579 (ASI-SS), 0.445 (GY-WW), and 0.368 (GY-SS). This leads to the last step, the result will be compared with the existing research which is using BL, BLUP, and RKHS methods in [9]. The comparison results can be seen in Table 1.

**Table 1.**Cross-validation (cv) correlation between predicted and observed phenotypes

| Trait–environment | Model[a] | | | |
|---|---|---|---|---|
| | RKHS | BL | BLUP | DBN |
| MFL – WW | 0.607 | 0.790 | —[b] | 0.295 |
| MFL – SS | 0.674 | 0.778 | 0.464 | 0.325 |
| FFL – WW | 0.588 | 0.781 | —[b] | 0.270 |
| FFL – SS | 0.648 | 0.774 | 0.521 | 0.370 |
| ASI – WW | 0.547 | 0.513 | 0.469 | **0.559** |
| ASI – SS | 0.572 | 0.517 | 0.481 | **0.579** |
| GY – WW | 0.514 | 0.525 | 0.515 | 0.445 |
| GY – SS | 0.453 | 0.415 | 0.442 | 0.383 |

Four models were fitted to each trait (FFL, MFL, ASI, and GY) and environment (SS, severe drought stress; WW, well watered) combination.

[a] Models were molecular marker (SNPs) using reproducing kernel Hilbert space (RKHS) regression, Bayesian LASSO (BL), best linear unbiased predictor (BLUP), and our proposed method deep belief network (DBN).

[b] BLUP were not computed because the estimated genetic variances were negligible [9].

Generally, DBN outperformed than other method and similarly or better than RKHS for anthesis-silking interval trait. DBN get a poor result for female and male flowering traits. This is because DBN is nonparametric method that build non-additive model for GP. Reference [15] provide evidence suggesting that female and male flowering traits in maize are additive traits. Therefore, one could expect the DBN to perform well in traits where epitasis plays a central role. RKHS is semiparametric method that may be able to capture epistatic interactions better than parametric method.Therefore, we can conclude that anthesis-silking trait in maize is a non-additive trait.

## 4. Conclusion

This research has succesfully modified and implemented DBN method for develop GP models. We also have compared the precision of DBN with other method that mostly used in GS. DBN outperforms than other methods, kernel Hilbert space (RKHS) regression, Bayesian LASSO (BL), best linear unbiased predictor (BLUP),in trait that we expect as non-additive traits. In this study, we have not done hyper-parameter optimization for training the DBN. We belief that DBN performance can be better with hyper-parameter optimization and apply advance method to prevent overfitting in neural network models.

## References

[1] Jannink J L, Lorenz A J and Iwata H 2010 *Brief. Funct. Genomics.* **9** 166-77
[2] Campos G D L, Naya H, Gianola D, Crossa J, Legarra A, Manfredi E, Weigel K and Cotes J M. 2009 *Genetics* **182** 375-85.
[3] Meuwissen T H, Hayes B J and Goddard M E 2001 *Genetics* **157** 1819–29.
[4] Ehret A, Hochstuhl D, Gianola D and Thaller G 2015 *Genet. Sel. Evol.* **47** 22.
[5] Ghahramani Z, Welling M, Cortes C, Lawrence N D and Weinberger K Q 2014 *Proc. Advances in Neural Information Processing Systems* (Montreal: NIPS Conference)
[6] Sainath T N, Mohammed A, Kingsbury B and Ramabhadran B 2013 *IEEE International Conference on Acoustics, Speech and Signal Processing* (ICASSP) May 26-31 (Vancouver: IEEE)
[7] Chen Q, Dong-Dong L, Shao-Long C, Yu-Ping W 2015 *Proceedings of the 2015 IEEE International Conference on Bioinformatics and Biomedicine* (BIBM) (USA :Washington DC)
[8] Hinton G E, Osindero S and Teh Y W 2006 *Neural Comput.* **18** 1527-54.
[9] Crossa J, De Los Campos G, Pérez P, Gianola D, Burgueño J, Araus J L, Makumbi D, Singh R P, Dreisigacker S, Yan J, Arief V, Banziger M and Braun H J 2010 *Genetics* **186** 713-24
[10] Gianola D and Kaam J B C H M V 2008 *Genetics* **178** 2289-303
[11] Bengio Y 2009 *FNT in Machine Learning* **2** 1-127.
[12] Glorot X and Bengio Y 2010 *Proceedings of the Thirteenth International Conference on Artificial Intelligence and Statistics* (AISTATS-10) (Brookline: Microtome)
[13] Mu L, Tong Z, Yuqiang C, Smola A J 2014 *Proceedings of the 20th ACM SIGKDD International Conference on Knowledge Discovery and Data Mining* (New York: Association for Computing Machinery)
[14] Theano Development Team 2016 Theano: A Python framework for fast computation of mathematical expressions.; arXiv *e-prints*. abs/1605.02688.
[15] Buckler E S, Holland J B, Bradbury P J, Acharya C B, Brown P J, Browne C *et al* 2009 *Science* **325** 714-8

IOP Conf. Series: Journal of Physics: Conf. Series **835** (2017) 012004 doi:10.1088/1742-6596/835/1/012004

# The implementation of bit-parallelism for DNA sequence alignment

**Setyorini*, Kuspriyanto, D H Widyantoro and A Pancoro**

School of Electrical & Informatic Engineering, Bandung Institute of Technology, Bandung, Indonesia, [4]Schools of Life Science , TB Bandung Indonesia

E-mail:setyorinianwars@students.itb.ac.id

**Abstract**. Dynamic Programming (DP) remain the central algorithm of biological sequence alignment. Matching score computation is the most time-consuming process. Bit-parallelism is one of approximate string matching techniques that transform DP matrix cell unit processing into word unit (groups of cell). Bit-parallelism computate the scores column-wise. Adopting from word processing in computer system work, this technique promise reducing time in score computing process in DP matrix. In this paper, we implement bit-parallelism technique for DNA sequence alignment. Our bit-parallelism implementation haveless time for score computational process but still need improvement for thereconstruction process.

## 1. Introduction

Alignment is the important first step toward structural and functional analysis of newly determined sequences. Many new biological sequences are being generated and already existing in the databases. Functional and evolutionary inference of new biological sequence can be drawn by sequence comparison from that database. Sequence alignment does acomparison by searching for common character pattern and making residue-residue correspondence among related sequences. Pairwise sequence alignment is the process of making alignment between 2 protein sequences [1].

As a computational problem, sequence comparison is defined as approximate string matching problem. Approximate string matching is a technique finding pattern of string that match approximately. There are limited errors that allowed during matching process. There are several techniques to implement approximate string matching : Dynamic Programming (DP) [2], Automaton [3], Filtering [4] and Bit-Parallelism

DP is one of the classicalalgorithms to solve approximate string matching problem. Classical DP computation process increases exponentially along with the increasing of sequence size. Several algorithms proposed in an effort to make the computational process more efficient, without sacrificing the accuracy [5]. Bit-parallelism is the one of that improvement , that parallelize the proses of score computation in DP matrix.

The authors implement bit-parallelism for aligning the DNA sequence, which is originally proposed by Myer [6]. Myer proposed bit-parallelism algorithm to speed-up DP score computation by column-wise computation. Alignment result obtained by reconstruction algorithm proposed by Hyrro [7] slightly modified to accelerate the computing process.

* Corresponding author

This paper is organized as follows : in Section 2 we describe sequence alignment as approximate string matching problem, that bit-parallelism DP matrix as one approach. Explanation of bit-parallelism and alignment construction mechanism will be obtained in Section 3. We present implementation and result discussion in Section 4, and then the conclusion in Section 5.

## 2. Sequence Alignment as a Problem of Approximate String Matching

Approximate string matching is defined formally by [5] as searching short pattern $P$ of length $m$, in a long text $T$ of length $n$, with up to $k$error. The output is all text position $j$, such that suffix $T \ldots j$ matches $P$ with at most k errors. Error defined as theminimum cost of sequence operations to transform pattern $P$ to text $T$. The cost of sequence operations is sum of individual operations. The sequence operations are aninsertion, deletion, and substitution. The error also is known as adistance function. Commonly used distance functions are : Levenshtein/Edit Distance, Hamming Distance, Episode Distance and Longest Common of subsequences Distance.

Sequence alignment consists of two tasks : searching the best score of a common pattern of sequence residues and constructing residue correspondence among them. The common pattern is selected by similarity level. This needs a scoring mechanism to determine the quality of similarity. Scoring scheme defines usage rules of edit operations into three assessment criteria, the score for a match, mismatch, and insertion.Gap character is used to represent insertion and deletion. Inserting and removing naturally occur in evolution. As approximate string matching problem sequence alignment defined as searching optimal total score of edit operation from all possible sequence residues comparations.

DP algorithm computed the edit distance. This algorithm is flexible to handle different distance function, but it is not very efficient. DP matrix increases exponentially as the sequence size increases. Comparing more than 2 sequences is feasible, but not reliable on it computation.

## 3. Sequence Alignment using Bit-parallelism

### *3.1. DP Score Computation*

SomDynamic Programming (DP) computation for solving pairwise sequence alignment problemwill generate a matching score matrix. Score each cell matrix is decided by it adjacency. Each cell has 3 adjacency, upper side, diagonal side and left side.

Given seq$S_1$ (length $m$) and seq$S_2$(length $n$ ), DP algorithm will generate a matrix $C(m+1, n+1)$. The matrix initializes,and compute the score with this the well-known recurrence :

**Recurrence 1.**

$$C_{0,j} = j$$
$$C_{i,j} = if(S1_i = S2_j)\ then\ C_{i-1,j-1}$$
$$else\ \ \min(C_{i-1,j}, C_{i-1,j-1}, C_{i,j-1})$$

Each score represents the cumulative distance until the i-thcharacter of $S_1$ and j-th character of $S_2$. There is illustration for DP matrix computation :

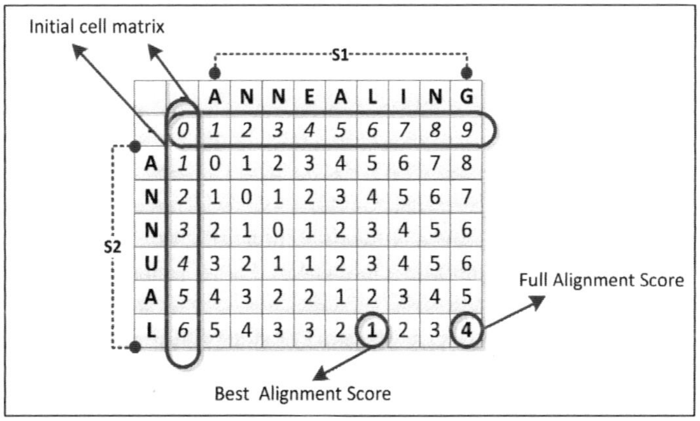


**Figure 1.** Example of dynamic programming matrix computation.

From matrix above, (Figure 1) the last line matrix can be seen as thefinal score. Scores for optimal full/global alignment is on the bottom right corner. Score represent the length of distance or differences between the seq$S_1$ and seq$S_2$. In this case, the best score is 1, because it does not belong to bottom right corner, it generates local alignment. All alignment result can be generated by tracingbackfrom each last line matrix score.

The example of full alignment results shown by Figure 2. Alignment result gives 5 matches, a mismatch, and 3 gap insertion. Matches/mismatches condition occur when the arrow of trace-back goes diagonally, the gap for string "ANNUAL " when the arrow goes vertically and gap for string "ANNEALING" when the arrow goes horizontally.

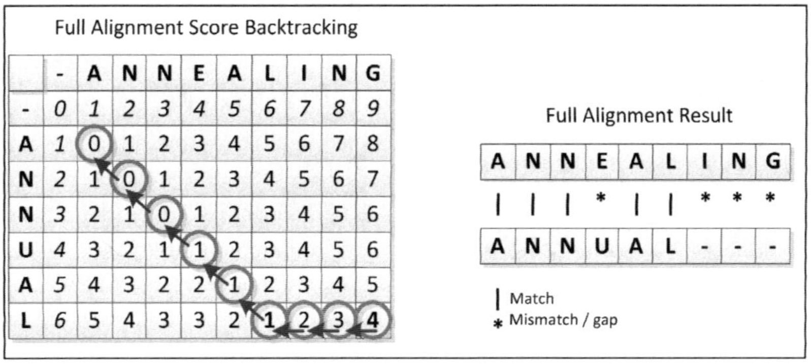


**Figure 2.** Example of alignment reconstruction from Dynamic Programming Matrix.

The idea bit-parallelism aimto save computing operations by converting the computing unit matrix cells from each cell into a collection of cells. Set of cells treated as new data unit such as the concept of word in the computer system. With this concept, the computing score DP replaced by computing the DP matrix cell adjacency. This difference value is then formed into a boolean value vector.

The adjacency of $C(i,j)$ cell matrix is shown in Figure 3.

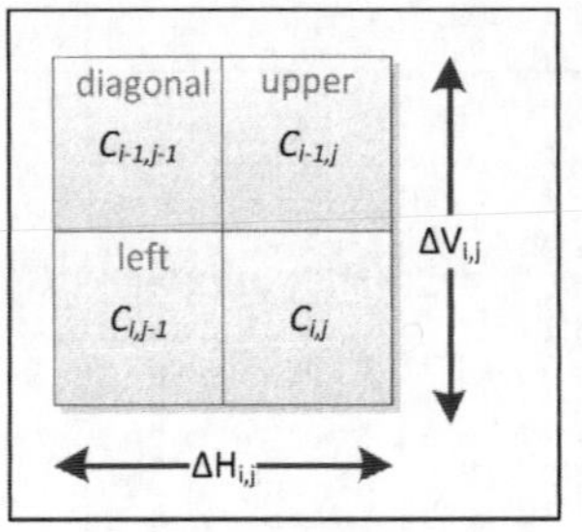


**Figure 3.** Myer adjacency DP cell vectors values.

There are 3 different kind of adjacency, vertically, horizizontally and diagonally Figure 3. With Recurence 1 the range of the adjacency values are represented as follows :

$$\Delta h_{i,j} = C_{i,j} - C_{i,j-1} \in \{-1,0,+1\} \quad (1)$$
$$\Delta v_{i,j} = C_{i,j} - C_{i-1,j} \in \{-1,0,+1\} \quad (2)$$
$$\Delta d_{i,j} = C_{i,j} - C_{i-1,j-1} \in \{0,+1\} \quad (3)$$

Two vectors required representing three possibility values in horizontal/vertical difference into boolean form. So the vectors formed as follows :

$$VP_j\ vector\ for\ \Delta V_{1..m,j} = +1$$
$$VN_j\ vector\ for\ \Delta V_{1..m,j} = -1$$

$$HP_j\ vector\ for\ \Delta H_{1..m,j} = +1$$
$$HN_j\ vector\ for\ \Delta H_{1..m,j} = -1$$
$$D0_j\ vector\ for\ \Delta V_{1..m,j} = +1$$

The next vector can be operated with the concept of bits. The bit operator involve are "& " for and, operator "+" for add operator, "|" for or operator, and "∧" for xor operator.

**Recurence 2 :**

$$D0_j \leftarrow (((PM_j \& VP_{j-1}) + VP_{j-1}) \wedge VP_{j-1}) | VP_{j-1} | VN_{j-1}$$
$$HP_j \leftarrow VN_{j-1} | (\sim(D0_j | VP_{j-1}))$$
$$HN_j \leftarrow D0_j | VP_{j-1}$$
$$if\ HP_j \& 10^{m-1} \neq 0\ then\ D[m,j] \leftarrow D[m,j] + 1$$
$$if\ HN_j \& 10^{m-1} \neq 0\ then\ D[m,j] \leftarrow D[m,j] - 1$$
$$VP_j \leftarrow (HN_j \ll 1) | \sim(D0_j | ((HP_j \ll 1) | 1))$$
$$VN_j \leftarrow D0_j \& ((HP_j \ll 1) | 1)$$

In Figure 4 is the example of generating the 5 vectors. A vector can be represented as an array of boolean values or as one/more integers values.

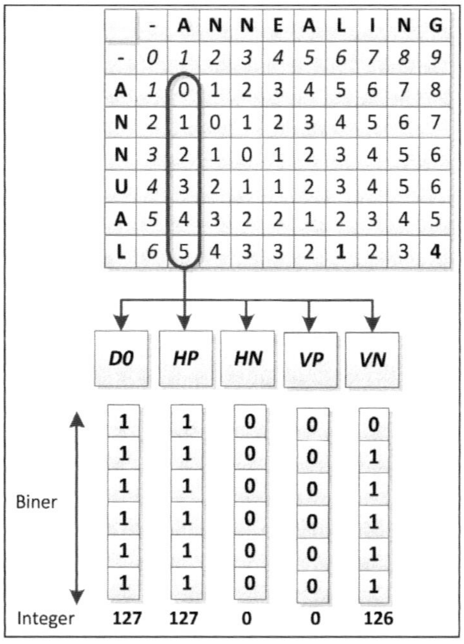


**Figure 4.** The example of constructing D0, HP, HN, VP and VN vectors.

### *3.2. Alignment Reconstruction*

Alignment reconstruction is about recovering alignment result from bit-parallelism vectors. Alignment reconstruction of DP matrix produced by running backtracking procedure. Neighboring cell value is checked to know from where the value is obtained. Recovering simple edit distance for LCS proposed by Iliopoulos and Pinzon[8], which permits a single operation to insert or delete a character.

Algorithm that handling more edit operation, substitution of character, is proposed by Heikki Hyyro [7]. This algorithm using vertical delta vector VN and VP (Figure 5). As backtracking in classical DP, this algorithm also start from the most right bottom. First, it checks if VP set, means the value come from the upper side, alignment result will be inserted gap for the pattern. But, if the VP is not set then it check VN. VN set means the value must become from the left side.

```
Alignment Reconstruction (DO,VP,VN)
i ←m, j ←n
while i> 0 and j>0 Do
if VP(i,j) ==1 then
      output (S1(i), '-')
      i ← i-1
else
if VN(i,j-1)== 1 then
        output ('-', S2(j))
else
        output (S1(i),S2(j))
        i ← i-1
      j ← j -1

whileI>0 do
    output (S1(i), '-'); i ←i-1
while j>0 do
    output ('-',S2(j)); j ←j -1
```

**Figure 5.** Hyrro algorithm for constructing alignment using vector VP & VN [7]

Using this algorithm we made a reconstruction matrix, during each column score computation processes. Next, is the principal to:

**Table 1**. Truth table of reconstruction matrix.

| VPj | VNj-1 | source cell |
|---|---|---|
| 1 | ϕ | 01 |
| 0 | 1 | 10 |
| 0 | 0 | 00 |

Use Table 1, truth table will produce output value for reconstruction matrix. There are three possibilities, 00 if the valuecomes from a diagonal direction (match/mismatch representation), 10 if come from above direction, and 11 if come from the left side (gap insertion). Thus, can be generated output function that produces output in relational logic from as follows;

$$Constrution\ Matrix = not(VP_j)VN_{j-1}.2^1 + VP.2^0 \quad (4)$$

Reconstructions algorithm have to maintain all iterative update of matrix VP and VN.

## 4. Sequence Alignment using Bit-parallelism

This paper applies the algorithms bit-parallelism and classical DP for computing the edit distance and alignment reconstruction. Alignment of the global alignment is intended. The system will require input as a pair of DNA Sequence in several sizes. A pair of sequence alignment is made in the same size. The DNA Sequence only contains 4 characters, as {'A', 'T', 'G', 'C'}.

In referring from all the pair sequence groups in the same size, Sequence alignment test result will be calculating from average of test parameters as follows: (1) identical pairwise residue, i.e. the percentage alignment results with the same character pairs, (2) pairwise mismatch residue, can be % of mismatching residue alignment and (3) the percentage of large number gap insertion. All test parameters are measured relative to initial sequence size

Alignment computational time is measured from two main processes, namely the computational score and reconstruction alignment results based on matrix score. Computing time score is the time to calculate all possible distance.

Datasets used in this study refer to data sets on the Loving research [9], which consists of 200 sequences with the length of each sequence is 100 residues. The data is then processed for generating some data size as follows:

**Table 2.** DNA data set.

| sequence size | # of sequence |
|---|---|
| 10 | 2000 |
| 20 | 1000 |
| 25 | 800 |
| 50 | 400 |

For Classic DP algorithm implementations adopts Needleman Wunch[10], with modifications to the weighting used, 0 for the conditions of the match and 1 for mismatch conditions and gap insertion. The weights used are the same with weights on the bit-parallelism algorithms Hyrro.Reconstruction algorithm adopted from the previous explanations on the Hyrro algorithm[7].

Alignment results are presented in the form of 3 string rows. The first and third line is a sequence of characters/residue from the sequence or the notation gap. The second line is the status of the correspondence between the residues. The symbol ' | ' to represent the match condition, the symbol ': ' for a mismatch, and the notation '-' to represent the gap insertion exist.

IOP Conf. Series: Journal of Physics: Conf. Series **835** (2017) 012004 doi:10.1088/1742-6596/835/1/012004

Figure 6 the following show average value for identical residue-pair percentage, and gap insertion percentage.

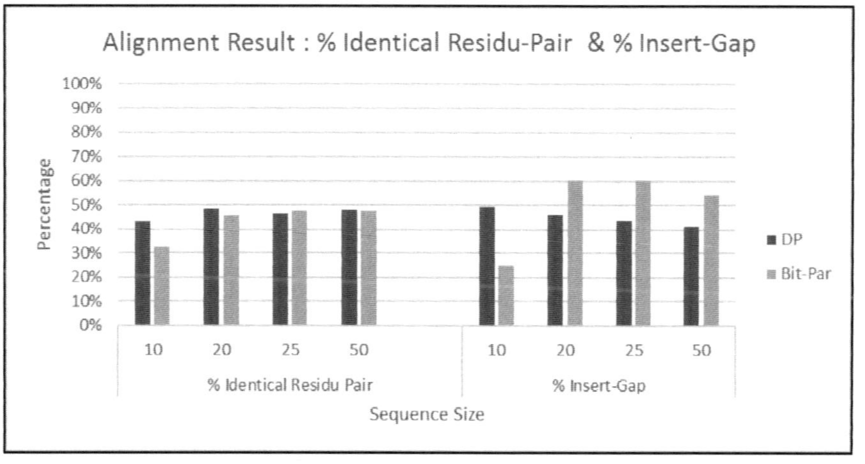


**Figure 6.** Sequence Alignment Result: average of procentage of match found & gap insertion.

Bit-parallelism got the similar result in the percentage of identical residu-pair. This result show that the algorithm still maintain the accuracy.

```
CACCAGACCAGCGCGGCTAGCGT-TT-TTGGGACT-GCGGACT-CT-CTGGCCGA-
|:  || |||||:|: | || || || :|:|:|:| ||:|:|| :| |:||::||
CT--AG-CCAGCTCT-C-AG-GTCTTCATCGCATTCGCAGCCTGGTGCCGGGAGAA
```

**Figure 7**. Best sequence alignment result for sequence size = 50

The best percentage identical residu-pair for sequence size 50 is 62% (Figure 7). However the algorithm still insert more gap than classical DP. Most of them are consecutives gaps (Figure 8) that occured at the beginning and the end of sequence.

```
-----CACCAGACCAGCGCGGCTAGCGTT-TTTGGGACTGCGGACTCTCTG-GCCG-A-
      |::|||:  | |:: ||:|:||| |||::: | ||| | |:||:| :::| |
TTGGT-AGGAGAT--G-GTA-CTGGAGTTGTTTCTC-C-GCG-A-TATCAGCATTGTAC

--------------CACCAGACCAGCGCGGCTAG-CGTTTTTGGGACTGCGGACTCTCTGGCCGA
              |    ||| | ||:|:|:|| |:|||  |:|: | ||:|:|:|| | |
CTCCTCCGGAGAGGC----GAC-A-CGTGCCCAGTCATTT--GCGT-T-CGAAGTTTC-G-C---
```

**Figure 8**. Examples of consecutives gap

Consecutives gap can be handled with the gap penalty technique, such as affine gap that combine contant and linear gap penalty. It uses 2 different weight for opening gap and extention gap. Bit-parallelism have no flexibility different weights is not directly applicable. It need different scoring technique. Loving [9] proposed general integer scoring for sequence alignment. Loving algorithm assume a integer score for match, mismatch and gap weight.

IOP Conf. Series: Journal of Physics: Conf. Series **835** (2017) 012004 doi:10.1088/1742-6596/835/1/012004

The different weights (Figure 9) also have impact on accuracy as the next case in DP technique :

```
match =0, mismatch,gap=1

TTGGTAGGA-G-A-T-GGTAC-TGGAGTTGTTTCTC-CGC-GATATCAGCATTGTAC
   :::::: : | : :|::: ::|::| :|::::| ::: ::|: ::: :::::|
---CACCAGACCAGCGCGGCTAGCGTTT-TTGGGACTGCGGACTC-TCT-GGCCGA-

identik = 8

match =0, mismatch=1,gap=2

TTGGTAGGAGA-TGGTACTGGAGTTGTTTCTCCG-CGATATCAGCATTGTAC
 ::::||:::| :|:::||:|:|||:||::::|: ||:::|| :|:::::::
-CACCAGACCAGCGCGGCTAGCGTTTTTGGGACTGCGGACTC-TCTGGCCGA

identik=18

match =0, mismatch=1,gap=5

TTGGTAGGAGATGGTACTGGAGTTGTTTCTCCGCGATATCAGCATTGTAC|
::::::::::::|:::||:|:|||:||::::|:::::|::::|:::::::
CACCAGACCAGCGCGGCTAGCGTTTTTGGGACTGCGGACTCTCTGGCCGA

identik = 12
```

**Figure 9.** Cases when different weights give impact on the sequence alignment accuracy

High accuracy occured when the composition of match, mismatch and gap weights give the minimum distance to the sequences.

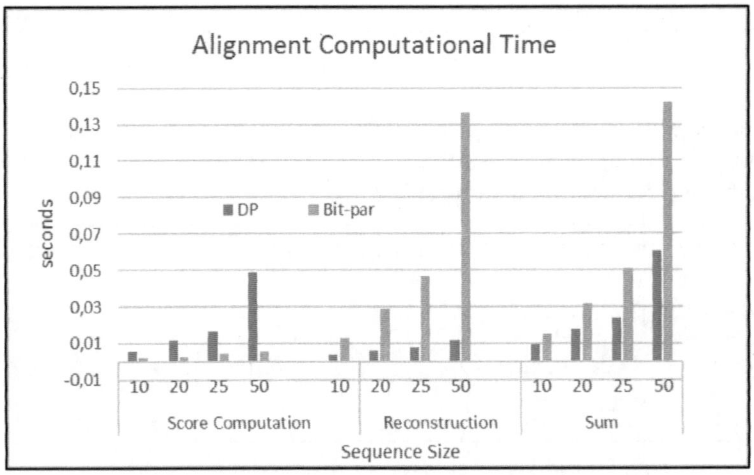


**Figure 10.** Sequence Alignment computation time

For computational time in Figure 10. show that bit-parallelism algorithm outperform at score computation phase, but still underperform at reconstruction phase. In our implementation need, reconstruction process consist of 2 phase, filling the direction matrix for backtracking and mapping the result. Matrix filling phase use the formula (1) did not consume much time, but not for iterative process in mapping phase.

## 5. Conclusion

The implementation of DNA sequence alignment using Bit-Parallelism technique produced similar accuracy as DP technique. Bit-Parallelism also increase the efficiency in score computation phase, however reconstruction phase still need to be improved. Implementation on more flexible weights

IOP Conf. Series: Journal of Physics: Conf. Series **835** (2017) 012004 doi:10.1088/1742-6596/835/1/012004

needed as future work to improved the accuracy and do more complex sequence alignment problem such on protein. This also open the posibility to implement the computation on hardware level.

**References**

[1] Rosenberg M S 2009 *Sequence alignment: methods, models, concepts, and strategies* (California: Univ of California Press)
[2] Xiong J 2006 *Essential bioinformatics* (Cambridge: Cambridge University Press)
[3] Wu S dan Manber U 1992 *Commun. ACM* **35** 83-91
[4] Wright A H 1994 *Softw. Pract. Exp.* **24** 337-362
[5] Ukkonen E 1992 *Theor. Comput. Sci.* **92** 191-211
[6] Smith T F and Waterman M S 1981 *Adv. Appl. Math.* **2** 482-489
[7] Smith T F, Waterman M S and Fitch W M 1981 *J. Mol. Evol.* **18** 38-46
[8] Sellers P H 1974 *SIAM J. Appl. Math.* **26** 787-793
[9] Needleman S B and Wunsch C D 1970 *J. Mol. Biol.* **48** 443-453
[10] Navarro G 2001 *ACM Comput. Surv. (CSUR)* **33** 31-88
[11] Myers G 1999 *Journal of the ACM (JACM)* **46** 395-415
[12] Masek E J and Paterson M S 1980 *J. Mol. Evol.* **20** 18-31
[13] Loving J, Hernandez Y and Benson G 2014 *Bioinformatics* **30** 3166-3173
[14] Landau G M and Vishkin U 1989 *J. Algorithm.* **10** 157-169
[15] Iliopoulos C S and Pinzon Y J 2001 *Proc of the 12th Australasian Workshop On Combinatorial Algorithms* (Australia)
[16] Iliopoulos C S and Pinzon Y J 2001 *Proc of the 12th Australasian Workshop On Combinatorial Algorithms* (Australia)
[17] Hyyrö H, Fredriksson K dan Navarro G 2005 *JEA* **10** 2-6
[18] Hyyrö H 2001 *Explaining and extending the bit-parallel approximate string matching algorithm of Myers* (Technical Report A-2001-10) (Tampere: University of Tampere)
[19] Hyyrö H dan Navarro G 2005 *Algorithmica* **41** 203-231
[20] Hyyrö H 2004 *Proc. 15th Australasian Workshop on Combinatorial Algorithms* (Australia)
[21] Hyyrö H 2004 *Proc. Prague Stringology Conference* '04 (Prague : Vydavatelstv´ı CVUT) pp.79-87
[22] Hyyrö H 2003 *Nord. J. Comput.,* **10** 29-39
[23] Gotoh O 1982 *J. Mol. Biol.* **162** 705-708
[24] Benson G, Hernandez Y dan Loving J 2014 *Bioinformatics* **30** 3166-3173
[25] Benson G, Hernandez Y dan Loving J 2014 *Bioinformatics* **30** 3166-3173
[26] Baeza Y R and Navarro G 1999 *Algorithmica* **23** 127-158.
[27] Baeza Y R, Barbosa E F dan Ziviani N 1993 *XIII Congress of the Brazilian Computer Society*, (Florian polis: Anais do SBC'93) pp 11-24

IOP Conf. Series: Journal of Physics: Conf. Series **835** (2017) 012005 doi:10.1088/1742-6596/835/1/012005

# Molecular dynamic simulation of Trastuzumab F(ab')$_2$ structure in corporation with HER2 as a theranostic agent of breast cancer

**S Hermanto[1,3*], M Yusuf[2], A Mutalib[2], and S Hudiyono[1]**
[1]Department of Chemistry, Faculty of Mathematics and Science, University of Indonesia, Depok, Indonesia
[2]Department of Chemistry, Faculty of Mathematics and Science, Padjadjaran University, Bandung, Indonesia
[3]Study Program of Chemistry, Faculty of Science and Technology, Syarif Hidayatullah State Islamic University, Jakarta, Indonesia

E-mail: hermantokimia@uinjkt.ac.id

**Abstract.** Trastuzumab as intact IgG are well researched for theranostic agent in HER2 overexpressed breast cancer. However, due to the relatively large of molecules it is slowly moved and weak penetration of the target cells. Fragmentation of trastzumab has been developed by pepsin cleavages to get the F(ab')$_2$ fragments. To observe the stability and accessibility of F(ab')$_2$ structure in corporation with HER2 (human epidermal growth factor receptor-2), the structure of antibody modeling had been developed with 1IGT as a template. Molecular dynamics (MD) of the F(ab')$_2$ structure simulation has been done in the aqueous phase with AMBER trajectories for 20 ns. Computational visualization by VMD (Visual Molecular Dynamics) were applied to identify binding site interaction details between trastuzumab F(ab')$_2$ and HER2 receptor. The results of MD simulations indicated that the fragmentation of trastuzumab F(ab')$_2$ did not change the structure and conformation of F(ab')$_2$ as a whole, especially in the CDR (Complementarity Determining Region) area. SASA (solvent accessibility surface area) analysis on lysine residues showed that formation of conjugate DOTA-F(ab')$_2$ predicted occur on outside of the CDR regions so its not interfered with binding affinity for the HER2 receptor. The molecular dynamic simulation of DOTA-F(ab')$_2$ with HER2 receptor in aqueous system generated $\Delta G_{binding}$ more highly (15.5066 kkal/mol) than positive control HER2-Fab (-45.1446 kkal/mol).

## 1. Introduction

The use of monoclonal antibody (mAb) in the diagnosis and therapy of cancer has been developed in two decades. Selectivity and specificity of antibodies which high enough against the target cells is one of

[*] Corresponding author

reason why this techniques has been chosen [1][2][3]. The mechanism of immunotherapy aimed directly at cancer cells, where there is an antigen or receptor site on cancer cells that are recognized by antibodies. The monoclonal antibodies capable of recognizing specific antigens on the surface of cancer cells. Antibody-antigen interaction typically occurs in CDR areas (*Complementary Determining Region).* One of the antibodies that has been commonly used for breast cancer therapy is Trastuzumab Mab (Herceptin) [4-6].

Trastuzumab, also known in commercial name Herceptin, currently used in a part of the immunotherapy of advanced breast cancers, expecially in extensive metastasis and in general, solid tumors over-expressing HER2 [7]. Different clinical trials confirm the efficiency of the antibody as an anticancer treatment. Trastuzumab binds to HER2-ECD domain. Interaction trastuzumab Fab to domain IV of the HER2-ECD categorizes this site as a possible target for anticancer therapies [5][8].

Application of trastuzumab in breast cancer therapy still has many weaknesses, due to its size is relatively large (approximately 150 kD), thus resulting in the molecule is less effective in reaching target cells (*low pharmacokinetics)* and weak penetration of target cells (*poor tumor penetration).* This phenomena potentially generate an antibody response to the resistance of host cells [9]. To overcome these limitations, an approach that can be done is to reduce the size of the antibody to become smaller fragments by enzymatic hydrolysis or genetic engineering. Enzymatic fragmentation antibodies much easier to do because it does not require complicated steps and costs are relatively large and easy to do. Fragment bivalent $F(ab')_2$ can be produced from the breakdown of a disulfide bridge at the carbonyl side of the cysteine residues by pepsin hydrolisis which will produce fFc fragments and $F(ab')_2$ [10-11].

In this research, enzymatic fragmentation of trastuzumab has been conducted using pepsin to hydrolyze part of the antibody fragment that was not a CDR region (Complementary Determining regions). From the *in silico* results of these fragmentation obtained $F(ab')_2$ fragments with affinity expected not to change compared with the intact antibody. To study the mechanism of DOTA-F(ab')2-trastuzumab interaction with HER2 receptor conducted studies by *in silico* approach with molecular modeling and molecular dynamic (MD) of $F(ab')_2$. The stability and accessibility of $F(ab')_2$ trastuzumab explored by molecular modeling of $F(ab')_2$ with 1IGT murine antibody as a template [12]. The structure modeling of $F(ab')_2$ results observed through molecular dynamics simulation in the aqueous phase with MD trajectories 20 ns. The Fab simulation used as a reference for the first phase conjugate formation of DOTA-$F(ab')_2$ which interacted to HER2 receptor and its affinity compared with native structure of complex-HER2 Fab (PDB file 1N8Z) [13]. The affinity interactions explored by MMGBSA between HER2 receptor and ligand $F(ab')_2$ respectively. *In silico* study of this immunoconjugates interaction expected to support the further development of a radiopharmaceutical over expressing HER2 positive breast cancer.

## 2. Computational Methods

### *2.1. Structure modeling and CDR analysis the fragment F(ab')$_2$*

Protein structure modeling and CDR analysis of $F(ab')_2$ fragment has been done by using antibody modeler in Biovia Discovery Studio 4.5 and Rosetta-based modeler antibody based on sequence data of trastuzumab IgG obtained from http://www.drugbank.ca/drugs / DB00072 [14]. Antibody modeling constructed with 1IGT as a template [15]. Amino acid sequences of $F(ab')_2$ heavy chain and light chain respectively combined and CDR analyzed through Rosetta software modeler antibody (antibody http://rosie.rosettacommons.org/) [16]. Type of intermolecular interaction visualized by Biovia Discovery Studio 4.5 and the value of RMSD from $F(ab')_2$ modeling determined by sequence alignment of the template structure. The validity of the $F(ab')_2$ structure were analyzed using Ramachandran plot Assessment (http: //www.RAMPAGE Ramachandran Plot Assessment.htm) and structure modeling results are stored in pdb format [12].

### *2.2. Molecular dynamic of F(ab')$_2$*

*2.2.1. Preparation and parameterization.* Structure modeling of F(ab')$_2$ results (in PDB file) refinement with several stages of energy minimization. Before the F(ab')$_2$ structure minimized prior modification of the structure by eliminating protons/H atom and change some amino acid residues of certain significant effect on the conformation of 3D protein and geometry optimization of molecules such as histidine (HID / HIE / HIP) and cysteine (CYS / CYX). Modified structured of F(ab')2 is saved in a new file name (fab2_ready.pdb). Furthermore pdb file run on Leap program (using AMBER 14) with force fields ff14SB mode with the grounds parameterization topology and coordinate files appropriate for protein structure. Modified structure checked and neutralized by the addition of counter ions (10 $Na^+$ units) and solvated with water molecules using solvatebox fab TIP3PBOX10. F(ab')$_2$ solvation structure stored in a new file (fab2_ready_sol.pdb). Before the salvation parameters are stored, the parameters included in the molecular ion water by: loadamberparams frcmod.ionsjc_tip3p and parameterization file is stored in the form: saveamberparm c fab2_ready_sol.prmtop fab2_ready_sol.inpcrd [17].

*2.2.2. Minimization.* Minimization in the SANDER module done by using two energy minimization functions, *steepes descent* and *conjugate gradient.* Stepees descent mode in the early stages minimization in which all atoms contained in the system with a constant force field restrain 100 kcal /mol except for water molecules and counter ions, carried 1,000 minimization stage. The second minimization is done by conjugate gradient mode where's all done restrain protein molecules except for a hydrogen atom with an energy restrain 5 kcal $mol^{-1}$ as many as 2000 stage. Minimization of the third stage is done by providing harmonic restrain (protein backbone) with a 2000 phase minimization. The last stage minimization all the molecules is gradually unrestrain by 5000 stages minimization to minimize steric repulsion between atoms [18-19].

*2.2.3. Heating.* The system heated gradually until it reaches physiological body temperature (310$^o$K) with phases (0-100K: 100-200K: 200-310K, respectively 20ps) in a state NVT using harmonic restraints on the main frame of the atoms (backbone).

*2.2.4. Equilibrium.* After the heating process, the system reaches equilibrium at a temperature of physiological (310$^o$K) with weak restrain the protein, further equilibration in conditions of NTP by removing restrain gradually and move at a constant pressure with phase equilibration each for 100 ps, until the final stage where the molecule not restrained.

*2.2.5. MD production.* Production trajectory run at NTP conditions. Temperature was controlled by Langevin thermostat with collision frequency 1 ps-1 and pressure controlled with Beredsen barostat with coupling constants 1 ps and a target pressure of 1 bar. Non bond cut-off value of 10 A is used and the electrostatic bond length is measured by Particle Mesh Ewald. Time step used during production is 2 fs and the time scale (running trajectory 20 ns), all hydrogen atoms are given constraint by using SHAKE algorithm. Short range repulsion-dispersion interactions were smoothly truncated at 10 Å. The particle mesh Ewald (PME) method was used to calculate long range electrostatic interactions, with a maximum grid spacing of 2.5 Å and using fourth-order (cubic) interpolation for the fast Fourier transforms [20].

*2.2.6. Trajectory Analysis.* Trajectory analysis carried out by using ptraj module with the topology file simulation results (fab2ready_dry.prmtop). In order to be analyzed, all the coordinate files from MD simulation (in prod* .crd) combined with a script file (trajin/prod*.crd) into a trajout file (20ns_dry_fab2.binpos binpos). Furthermore, the combined results of the simulation file in the form file binpos calculated of RMSD and RMSF value (to identify fluctuations Fab structures and mobility of individual amino acid residues and analyzed the dynamics and flexibility of its 3D structure by using VMD (Visual Molecular Dynamics) 1.9.1. The crystal structure of Fab trastuzumab (1n8Z) used as a positive control to observe the conformational change, flexibility and accessibility during simulation trajectory with the timescale 20 ns [21].

*2.2.7. Prediction accessibility of DOTA conjugation to F (ab')$_2$.* Structure accessibility of F(ab')$_2$ binding DOTA conjugates determined by calculating the SASA (solvent accessibility of surface area) parameters at all outside lysine residues in the F(ab')2 molecule. The degrees of freedom of lysine residues obtained were plotted based on the SASA parameters by using trajectory file from simulation. The conjugation accessibility visualized through the program VMD (Visual Molecular Dynamics) 1.9.1.

*2.3. Molecular dynamic of HER2-Fab and HER2-DOTA-F(ab')$_2$.*
HER2-Fab complex structure (positive control) obtained from the crystal structure (1N8Z) by adding some missing residue using PRODAT databases based on homology to sequences obtained from NCBI. Modeling and analysis of complex structures (HER2-Fab) conducted in PDBsum (http://www.chemogenomix.com/chemogenomix/ PDBsum.html). The structure modeling of HER2-Fab crosscheck by the stereochemistry mapping with Ramachandran plot to observe the favorable conformation of each amino acid residues in the region are permitted (allowed region). Molecular dynamics simulations of Fab with HER2 receptor is done through the program AMBER 14 in a state of solvated complex water at pH 7 and physiological temperature (37°C). The simulation results of HER2-Fab complex was observed through cpptraj program by plotting RMSD value over time and comparing the fluctualization residues on each structure (ligands and receptors). Inventory of hydrogen bonding performed on the CDR regions involved in the interaction with the receptor. Visualization distance and probability (occupancy) formation of hydrogen bonds between the donor and acceptor analyzed through cpptraj on each residue based on the simulation results MD trajectory file. The calculation of free energy value carried out by MMGBSA module where all frames conformation from MD simulation results (complex, ligand and receptor) were included in the calculation to obtain the free binding energy [21-22].

## 3. Results And Discussion

*3.1. Structure Modeling of F(ab')$_2$.*
Before the molecular dynamic simulation , structure modeling of F(ab')$_2$ fragments carried out by using antibody modeller in Biovia Discovery Studio 4.5. Molecular dynamic (MD) simulation refers to the observation of conformational before and after conjugated with pSCN-DOTA molecules and predict accessibility for binding conjugate based on the SASA (*solvent assesibility surface area)* analysis. The results of structure modeling of F(ab')$_2$ as Figure 1.

Based on antibody modeling of F(ab')$_2$ structure with 1IGT as a template, it was appears that the F(ab')$_2$ fragment have a large similarity index sequences (78 408%) after superimpose with the 1IGT template by the RMSD value 1.5389 Å (Figure 2)., The overlay results indicate that the structure modeling of the F(ab')$_2$ results have a similar conformation to the template.

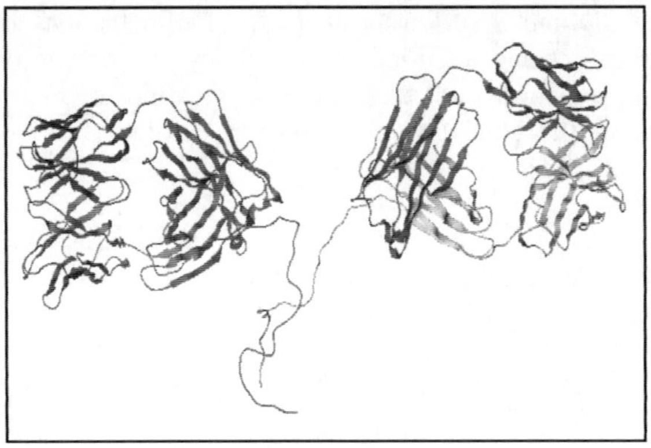

**Figure 1.** 3D structure of F(ab')2 fragments from antibody modeler (Biovia Discovery Studio 4.5)

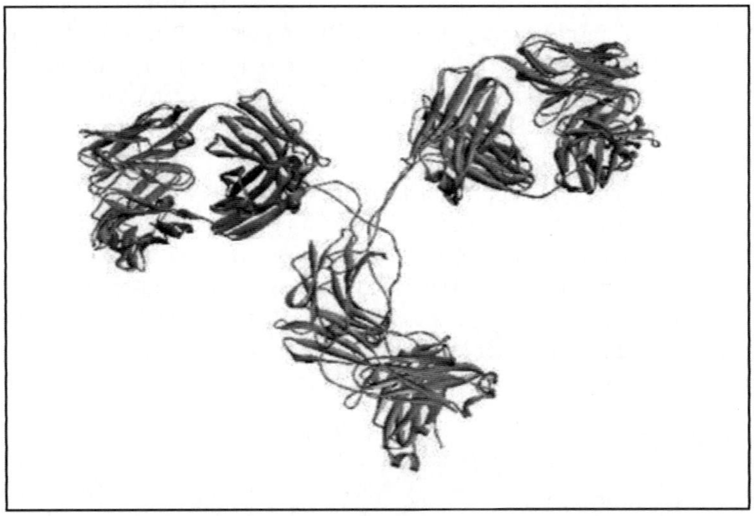

**Figure 2.** Superimpose of F(ab')$_2$ structure (blue) to the template structure 1IGT (yellow).

To ensure the validity of the structure modeling, the sequence alignment carried out to the template 1IGT in SCR areas (*structuraly conserve region)* and SVR (*structuraly variable region)* through multiple sequence alignment (MSA) by the method of Psi-BLAST Biovia software contained in Discovery Studio 4.5. This aligment to ensure that the structure of the F (ab')$_2$ used in accordance with sequences that are desired. The MSA can also be known that the area is maintained (conserved region) of the amino acid sequence fragments including the CDR (*Complementary Determining Region).* The result of the sequence alignment F(ab')$_2$ tratuzumab are as follows:

In the sequence alignment result (Figure 3) shows that the amino acid sequence of F(ab')$_2$ are retained (residues 1-220) naturaly conserve with relatively high homology especially in the CDR area (1-100). Although both include the type of F(ab')$_2$ trastuzumab and 1IGT (immunoglobulin) relative differ mainly in the area of non-CDR because of 1IGT is a chimeric monoclonal antibody types while F(ab')$_2$ is a type of humanized antibody. Chimeric monoclonal antibody is a combination of all parts of the Fv antibody mice with human antibody Fc (30% mouse antibody). Meanwhile, the humanized antibody (10% mouse antibody), did not take up the whole of a mouse antibody Fv part, but only part of the CDR only [23].

On the other hand, the mapping of the secondary structure of F(ab')$_2$ by using Ramachandran Plot Assessment (http: //www.RAMPAGE Ramachandran Plot Assessment.htm) (Figure 4) shows the stereochemical data structure was thermodynamically favorable where more 94% of the amino acid residues located in areas favorable and allowed regions and only 0.6% of residues located outside the region are allowed (disallowed region).

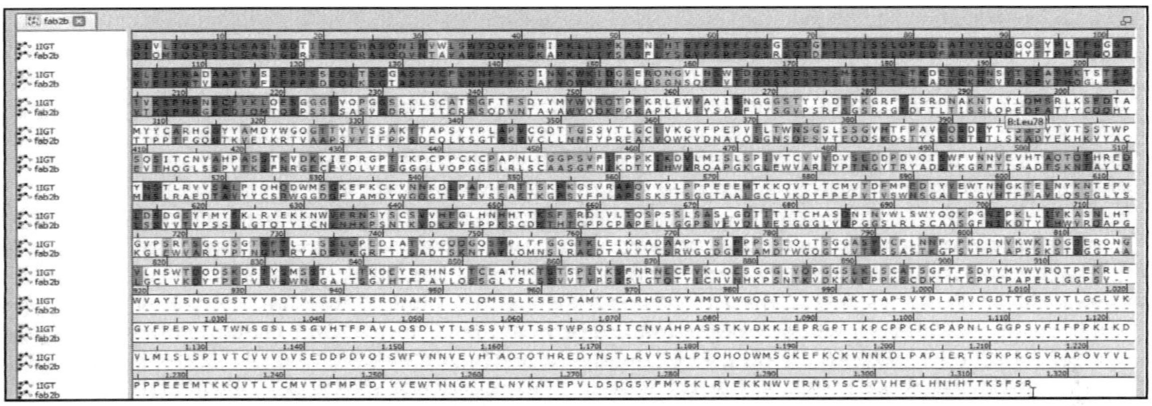

**Figure 3.** The multiple sequence alignment of F(ab')2 trastuzumab with 1IGT as template

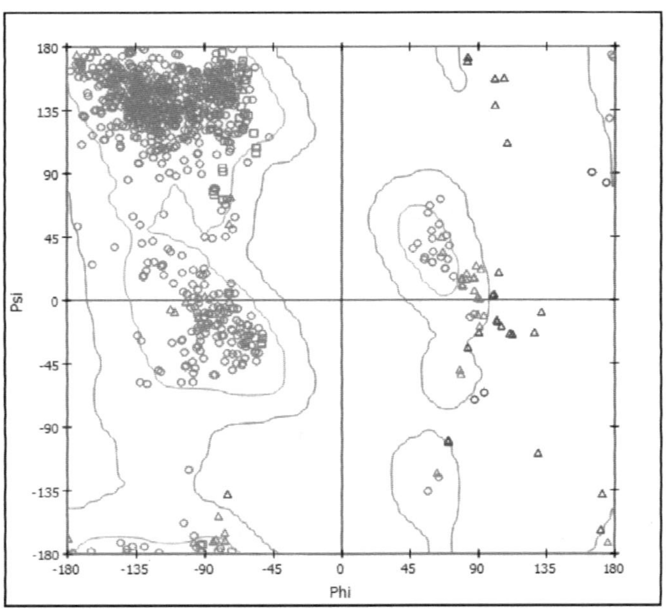


**Figure 4.** *Ramachandran Plot assessment* of F(ab')$_2$

Determination of CDR regions is done by using Rosetta antibody modeler [16], where the analysis results for each CDR sequences of the light chain and heavy chain are as follows (Figure 5).

**FASTA sequence of FV Light Chain (VL) with CDR L1, CDR L2, CDR L3 regions:**
DIQMTQSPSSLSASVGDRVTITC RASQDVNTAVA YQQKPGKAPKLLIY SASFLYS
GVPSRFSGSRSGTDFTLTISSLQPEDFATYYC QQHYTTPPT FGQGTKVEIKRT

**FASTA sequence of Heavy Chain (VH) with CDR H1, CDR H2, CDR H3 regions:**
EVQLVESGGGLVQPGGSLRLSCAAS GFNIKDTYIHWVRQAPGKGLEW VA
RIYPTNGYTRYADSVKGRFTISADTSKNTAYLQMNSLRAEDTAVY YCSR WGGDGFYAMDY
WGQGTLVTVSSA

**Figure 5.** CDR regions of Trastuzumab F(ab')$_2$ Light Chain and Heavy Chain

Based on analysis of CDR, obtained that CDR1 light chain regions of F(ab ')2 consist of 11 amino acid residues (residues 24-34), CDR2 region (residues 50-56) and CDR3 (residues 89-97) consists of 7 and 9 amino acid residues. In the heavy chain, CDR1 region consisting of 10 amino acid residues (residues 26-35) and CDR2 consists of 17 amino acid residues (residues 50-66) and CDR3 consists of 11 amino acid residues (residues 100-110). Each region is a part of the interface that is likely to interact strongly with domain area IV HER2 receptor. As predicted crystal structure of complex HER2-Fab, where residue contact amino acid Fab located at residue Trp99 that will interact with Phe573 through hydrophobic interactions, Asn30, Tyr92, Thr94, Arg50 and Gly103 to Gln602, Lys569, Asp560, Glu558, Asp560 and Lys593 through the interaction of hydrogen bonding and ionic interactions in the Asp560 residues Arg59 and Arg50 with Glu558 and Asp560. However, interaction need to be explored further through molecular dynamic simulation approach by mapping the interactions between the HER2 receptor structure that has been modeled Fab (PDB ID. 1N8Z) [13].

Accessibility and flexibility of the F(ab')2 structure were analyzed by observing changes in the value of RMSD and RMSF on the backbone structure of F(ab')$_2$ during the simulation period with 20 ns timescale at NTP conditions (temperature and constant pressure). The value of RMSD and RMSF is calculated based on the trajectory files that have been incorporated in the module cpptraj. RMSD (*root mean sequare deviation)* showed deviation in 3D structure during the simulation process F (ab')2 as a function of the coordinates of the starting position. On another hand RMSF (*root mean square Fluctuation)* is a measurement of the average mobility of atoms / residues.

### *3.2. Flexibility structure of F(ab')$_2$*

Based on RMSD fluctualization (Figure 6), it appears that the conformation of the 3D structure of F(ab')$_2$ relatively stable even changes in the value of RMSD relatively large (5-20A). RMSD changes in F(ab')$_2$ in total residue due to the flexibility of the hinge region were quite free as a consequence of antibody fragment cuts off. Conversely, if compared with the value of RMSD in region 1 and region 2, which contains CDR regions, the structure and conformation of F (ab')$_2$ is relatively more stable (RMSD <2A). The stability of the F (ab')$_2$ structure is likely influenced by the formation of intermolecular disulfide bridges, hydrogen bonding and electrostatic interaction between aspartic acid residues, glutamic acid and lysine. Flexibility of F(ab')$_2$ structure in the hinge region became the reason why the structure intact antibodies are generally difficult to be crystallized [21][24].

Flexibility and structure dynamics of F (ab')$_2$ at the residue levels can be explored more detail through the analysis of RMSF where mobility and flexibility of each residue is determined on the region 1 and region 2 area by comparing fluxtuality with the Fab structures (positive control). Fab structure derived from complex structure of HER2-Fab crystals (1N8Z) [13][21]. Based on the results of RMSF seen that almost all amino acid residues F(ab')$_2$ has relatively small fluxtuality when compared with the structure of the Fab with RMSF values below 2 except for a few specific residues (Figure 7).

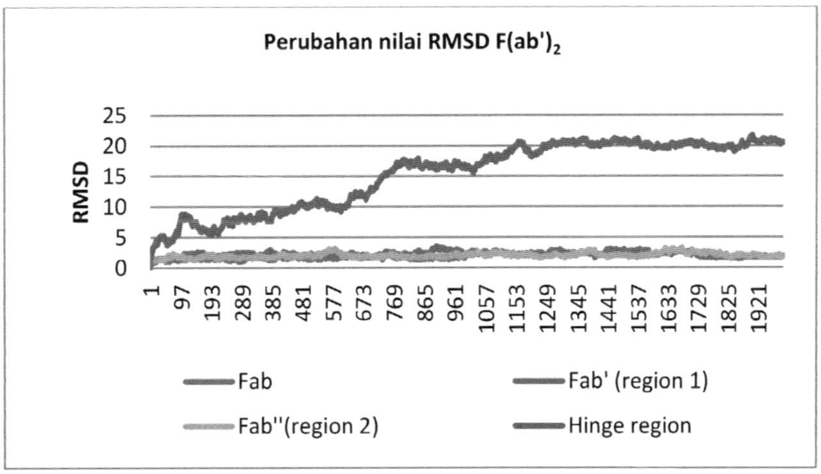


**Figure 6.** Flexibility of F(ab')2 from MD simulation (*time evolution plot 20 ns)*

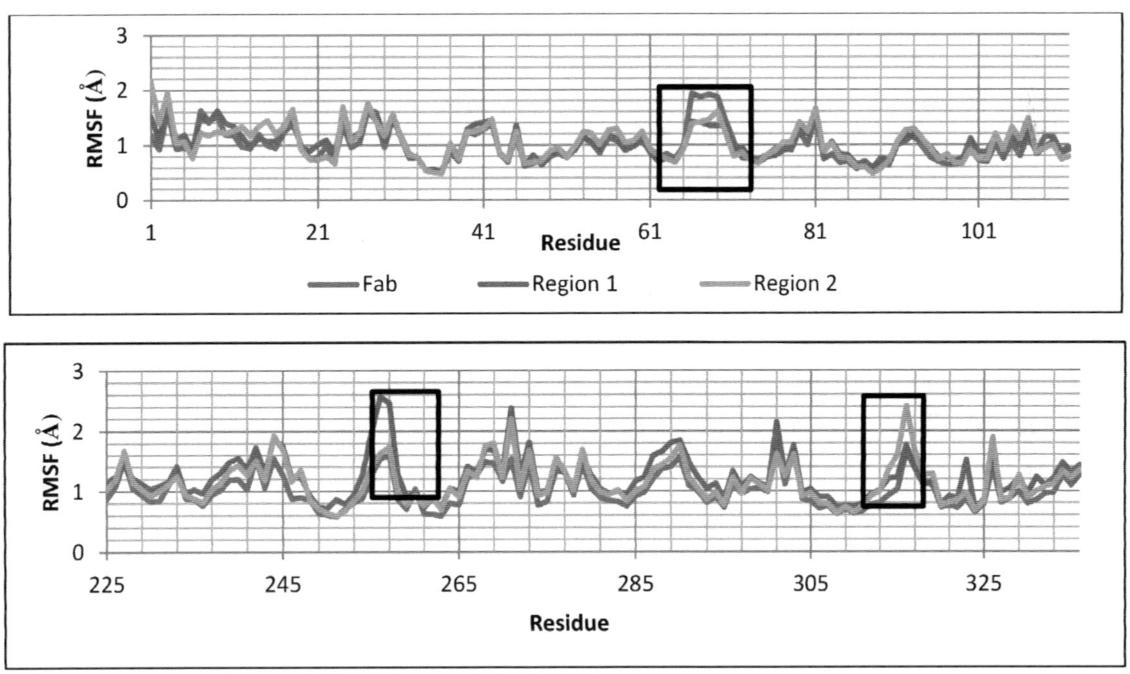


**Figure 7.** Amino acid residues fluxtuality of F(ab')$_2$

*3.3. Accessibility of F(ab ')2 and the conjugate mechanism of formation of DOTA-F(ab')$_2$*

Degrees of freedom of lysine residues and its accessibility to the pSCN-Bn-DOTA is determined by calculating the SASA on each lysine residue. From these calculations can predict where the all the most potential lysine residues exposed to the solvent so that thermodynamically allows for the conjugation reaction. SASA analysis of lysine residue results shown in Figure 8.

Based on the SASA analysis, there are three lysine residues on each arm of the F(ab')$_2$ structure which has high exposure to solvents, ie. Lys169 (residual light chain / non-CDR), Lys190 (residual light chain / non-CDR ) and Lys208 (residue of heavy chain / non-CDR) (Figure 8). The structure of the F(ab')$_2$ consists of two arms (Fab' and Fab ''), so there are 6 lysine residues that have a high enough exposure and are outside the CDR regions [25].

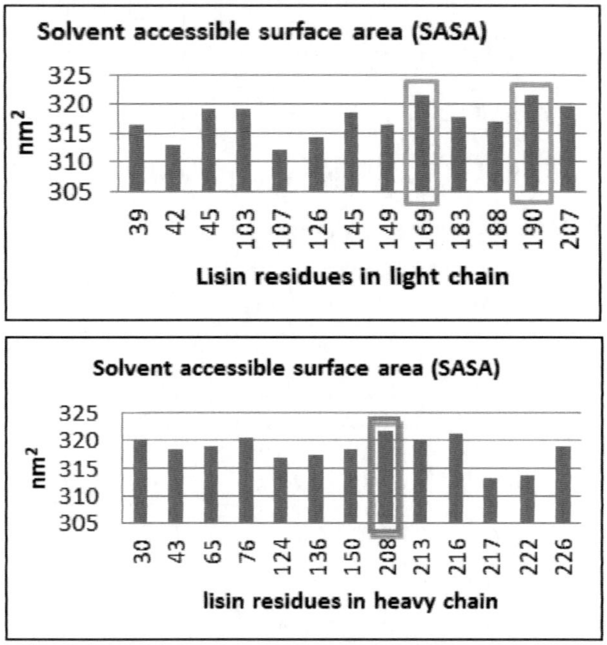


**Figure 8.** SASA Analysis of F(ab')2 (VMD 1.9.1)

*3.4. HER2-Fab molecular dynamic simulations in the water phase*

Molecular dynamic simulations of HER2-Fab conducted for 20 ns on NTP state (pressure and temperature constant). This simulation was performed with the same timescale as the simulation F(ab')$_2$ in a free state. The molecular dynamic simulation results show the RMSD value change as follows (Figure 9):

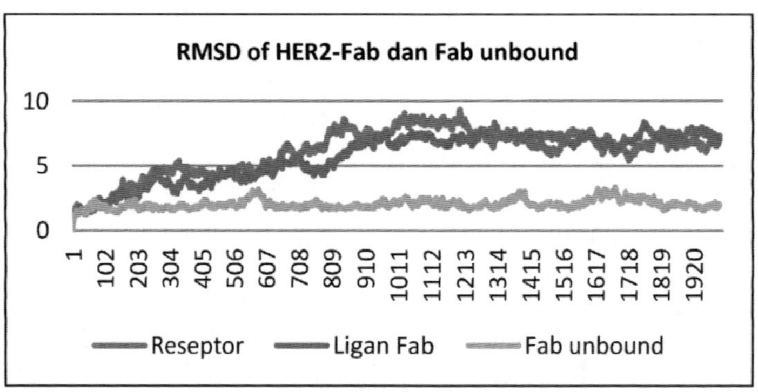


**Figure 9.** Flexibility of RMSD HER2-Fab and free Fab

Based on the RMSD value (Figure 9), HER2-Fab complex structure is relatively more flexible with RMSD value larger than the Fab structures in a free state / not bound to the receptor. Flexibility HER2-Fab complex structure likely influenced by the adjustment phase during the formation of complex systems and their interaction loop in domain II of HER2 with CH region (heavy chain) of Fab molecules. This was confirmed by a shift in conformation observed during the simulation period, as shown in Figure 10.

Conformational change of HER2-Fab complex in the water phase (during 1-10 ns) is an adjustment of induced fit effect between the Fab molecules (ligands) with HER2 (receptors) where local loop (domain II) HER2 change the orientation of approaching the CH molecule ligand complex Fab resulting in more stable interactions. HER2-Fab complex formation at the interface region is mediated by the presence of non-covalent interactions between residue found on the contact area with several Fab CDR amino acid residues in the interface (Figure 10). These interactions include intermolecular hydrogen bonding interactions, ionic interactions (salt bridges) and hydrophobic interactions (phi-phi aromatic). The third type of interaction is instrumental in the adjustment phase of the system so as to produce a stable compound [25-26].

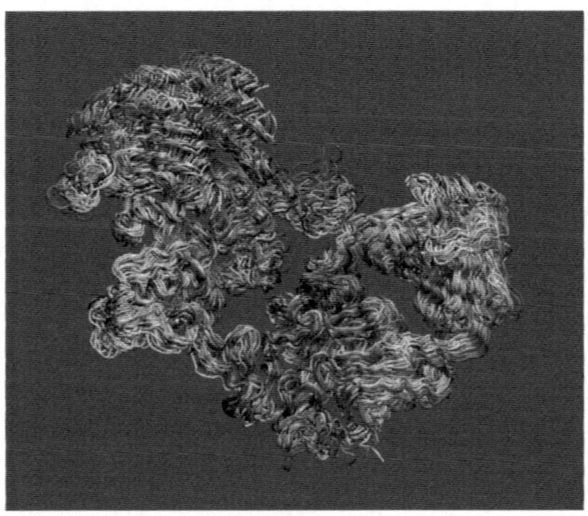

**Figure 10.** Conformational change of HER2-Fab complex during the simulation period (timesteps 1-999 (red to white), 1000-1999 (white to blue).

Through the calculation of the value of binding energy by using MM-GBSA (Poisson-Boltzmann surface area) values obtained for $\Delta G_{binding}$ -45.1446 kcal / mol. If compared with the results of the wet experiment analysis of $\Delta G_{binding}$ (12.4-14.10 kcal / mol), the value of the free energy of binding ligands relatively larger than the experimental values, It is likely influenced by the value of the entropy of the system in which the in silico simulation calculations entropy values are not fully included in the calculation [13][26].

Molecular dynamic simulation of the HER2 receptor with DOTA-F(ab')$_2$ prepared by the same procedures as the positive control (HER2-Fab). The binding energy result of complex interactions between HER2 and DOTA-F(ab')$_2$ calcultaed by MM-GBSA module generates the binding free energy ($\Delta G_{binding}$) amounted to 15.5066 kcal/mol. When compared with receptorligand binding affinity in the stimulation of HER2-Fab (positive control), the value of this $\Delta G_{binding}$ more positive, it means the relative binding affinity decreases. This is probably due to the presence of DOTA molecules in ligand structure F(ab')$_2$ which indirectly affect the stability of the complex conformational HER2-F(ab')$_2$ thus becomes a lower binding affinity.

## 4. Conclusions

The results of the molecular dynamics simulation F (ab ') indicates that fragmentation trastuzumab Fc molecules, the relative does not damage the structure and conformation of $F(ab')_2$ as a whole, especially in CDR areas that do not degrade binding affinity to HER2 receptors. The results of the SASA analysis on lysine residues of $F(ab')_2$ show that the reaction of formation of conjugate DOTA-F $(ab')_2$ is occured outside of the CDR regions so it does interfere with binding affinity for the receptor HER2. The results of the molecular dynamics simulation DOTAF$(ab')_2$-HER2 in the aqueous phase to produce a relatively larger $\Delta G_{binding}$ (15.5066 kcal / mol) compared with HER2-Fab complex (-45.1446 kcal / mol) or in other words the binding affinity decreases.

## References

[1] Khawli LA, Biela BH, Hu PS, and Epstein AL 2002 *Hybrid Hybridomics* **21** 11-8
[2] Boland WK and Bebb G 2009 *Expert Opin. Biol. Ther.* Sep; **9**(9):1199-206
[3] Gemmete JJ and Mukherji SK. 2011 *AJNR Am. J, Neuroradiol.* **32** 1373–4
[4] Phillips GDL, Li G, Dugger DL, Crocker LM, Parsons KL, Mai E, Blättler WA, Lambert JM, Chari RVJ, Lutz RJ, Wong WLT, Jacobson FS, Koeppen H, Schwall RH, and Sliwkowski MX 2008 *Cancer Res.* **68** 9280–90
[5] Fiszman GL, and Jasnis, MA 2011 *Int. J. Breast Cancer* **2011** 1-11
[6] Rasaneh S, Rajabi H, Babaei MH, Daha FJ, Salouti M. 2009 *Nucl. Med. Biol* **36** 363–9
[7] Pohlmann PR, Mayer IA and Mernaugh R 2009 *Clinical Cancer Res.* **15**(24) December 15
[8] Tai W, Mahato R, and Cheng K 2010 *Journal of Controlled Release* **146** 264–75
[9] Wong KJ, Baidoo KE, Nayak TK, Garmestani K, Martin WB, and Milenic, DE 2011 *EJNMMI Research* **1** 1
[10] Lamoyi E. 1986 *Meth. Enzymol.* **121** 652-663
[11] Lipman NS, Jackson LR, Trudel LJ, and Garcia FW 2005 *ILAR J* **46** 258-68
[12] Gonzalez JFF, Cruz VL, Ramos Z, and Salazar JM 2013 *J. Mol. Model* **19** 1227–36
[13] Cho HS, Mason K, Ramyar KX, Stanley AM, Gabelli SB, Denney WJ, and Leahy DJ 2003 *Nature* **421** 756-60
[14] Trastuzumab. (2014, November 14). Retrieved from www.drugbank.ca/drugs/DB00072
[15] Leach AR 1996 *Molecular Modeling: Principles and Applications* (Singapore: Longman)
[16] Sircar A, Kim ET, and Gray JJ 2009 *Nucleic Acids Res.* **37** W474–9
[17] Shang Y and Simmerling C 2012 *Molecular Dynamics Applied in Drug Discovery: The Case of HIV-1 Protease* In: Baron, R. Computational Drug Discovery and Design (New York: Springer)
[18] Berendsen HJC, Postma JPM, Van Gusteren WF, Dinola A, and Haak JR 1984 *J. Chem. Phys.* **81** 3684
[19] Rickaert JP, Ciccotti G and Berendsen HJC 1977 *J. Comput. Phys.* **23** 327-41
[20] Darden T, York D, and Pedersen L 1993 *J. Chem. Phys.* **98** 10089-92
[21] Brandt JP, Patapoff TW, and Aragon SR 2010 *Biophys J.* **99** 905–13
[22] Yusuf M 2015 *Molecular Dynamic Studies of The Effects of Neuraminidase Mutation on Oseltamivir Resistance* (PhD Thesis) (Penang: Universiti Sains Malaysia)
[23] Schmitz KR, and Ferguson KM 2009 *Exp. Cell. Res.* **315** 659 – 70
[24] Telesco SE and Radhakrishnan R 2009 *Biophys. J.* **96** 2321–34
[25] Fuentes G, Scaltriti M, Baselga J and Verma CS 2011 *Breast Cancer Res* **13(3)** R54 1-9
[26] Troise F, Cafaro V, Giancola C, D'Alessio G, and Lorenzo C 2008 *FEBS J.* **275** 4967–79

IOP Conf. Series: Journal of Physics: Conf. Series **835** (2017) 012006 doi:10.1088/1742-6596/835/1/012006

# Combining PSSM and physicochemical feature for protein structure prediction with support vector machine

**I Kurniawan, T Haryanto*, L S Hasibuan, M A Agmalaro**

Department of Computer Science, Bogor Agricultural University, Indonesia

E-mail: totoharyanto@apps.ipb.ac.id

**Abstract**. Protein is one of the giant biomolecules that act as the main component of the organism. Protein is formed from building blocks namely amino acids. Hierarchically, the structure of protein is divided into four levels: primary, secondary, tertiary, and quaternary structure. Protein secondary structure is formed by amino acid sequences that would form three-dimensional structures and have information about the tertiary structure and function of proteins. This study used 277,389 protein residue data from enzyme categories. Position-specific scoring matrix (PSSM) profile and physicochemical are used for features. This study developed support vector machine models to predict the protein secondary structure by recognizing patterns of amino acid sequences. The Q3 results showed that the best scores obtained are 93.16% from the dataset that has 260 features with the radial kernel. Combining PSSM and physicochemical feature additions can be used for prediction.

## 1. Introduction

Protein is one of the giant biomolecules that act as the main component of the organism. Protein is formed by amino acid sequence linked to each other by peptide bonds through the carbon chain so that to form complex structures. The process of protein formation process involves translation and transcription. The transcription process is writing the genetic code of deoxyribonucleic acid (DNA) into messenger ribonucleic acid (mRNA) to encode each nitrogenous bases of DNA into RNA in the form of nitrogenous bases. The translation process is translating each nitrogenous bases in mRNA into the amino acid sequence [1]

Hierarchically, proteins are divided into four levels, primary structure, secondary structure, tertiary structure, and tertiary structure. The primary structure is the sequence of amino acids that form a polypeptide chain. Secondary structure is a series of amino acids that form a three-dimensional structure alpha helix (H), beta-sheet (B), or coil (C) which is the result of amino acid sequences that bind to the peptide bond [1]. Tertiary structure is a combination of the secondary structure after folding process. The role of the protein can be known if it formed the tertiary structure in 3D. However, the tertiary structure can be determined if the structures previously known.

Conventionally, the structure of a protein can be detected by X-ray crystallography [2] and Nuclear Magnetic Resonance (NMR) [3]. This technique is used to find the primary structure and new structure of proteins and contribute to the validation of protein structures. But both these techniques take time and cost are relatively expensive. Thus, the computing-based approach is widely used to

* Corresponding author

predict the secondary structure of proteins. Protein secondary structure prediction is performed to find the 3D structure of proteins based on the primary structure of proteins. There are two methods of protein secondary structure prediction, the method of comparative modeling and modeling of de novo or ab initio. Comparative protein modeling to predict protein structure based on the structure of other proteins that are known, while the method of ab initio or de novo to determine the structure of proteins from primary sequence without comparing with other protein structure [4].

The basis of the protein secondary structure classification system is by recognizing pattern in the protein amino acid sequences. Many related types of research have been conducted to predict the secondary structure of this protein in order to get the best accuracy from previous studies. Huang and Chen [5] predict the secondary structure of proteins by using Support Vector Machine with Position Specific Scoring Matrix (PSSM) and four feature extraction consisting of conformation parameters, net charges, hydrophobic, and side chain mass. This research resulted in the accuracy of the value of Q3 score of 75.92%. Position based structure was proposed for secondary structure prediction with accuracy in ranging 87.6% up to 95.6% [6].

Support Vector Network had been introduced [7] and known as Support Vector Machine (SVM). This method was first presented at the Annual Workshop on Computational Learning Theory. The basic concept of SVM is a harmonious combination of computational theories that already exist such as margin hyperplane and kernel. SVM perform a technique to discover the function of separator that separates the two sets of data from two different classes. This method is a machine learning method that works on the principle of Structural Risk Minimization (SRM) with the goal of finding the best hyperplane that separates two classes in the input space. Basically, SVM working with the principle of linear classifier, then developed to work on nonlinear case by using kernel concept in high-dimensional workspace.

Neural Network (NN) and SVM with their variance are prominent technique for secondary structure prediction. Knowledge based and NN as one approach for secondary structure prediction [8], variance of NN, SOM and SOGR also proposed in [9] and JNet algorithm also proposed [10]. Mix modal of SVM proposed in [11] with PSI-BLAST for input feature. Even, improved SVM and Neural Network was introduced in [12] for protein secondary structure perediction.

According to the problems described in the background, problem formulation of this research is how to predict the secondary structure of proteins using SVM method and analyze the effect of using PSSM profile and physicochemical features on SVM-based model accuracy. Feature extraction on protein structure prediction is one of the main interest research focus. This study purposed to combine PSSM and physicochemical as feature extraction with 13 sliding window refers to [5] on protein secondary structure prediction. This research is expected to produce a good model to predict the secondary structure of proteins before predict the 3D structure of proteins in order to identify further structure and function of the protein. The assignemnt of secondary structure label derived from Dictionary of Secondary Structure of Proteins in short DSSP with enzim commision data from Protein Data Bank (PDB).

## 2. Methodology

This chapter is divided into eight sections includes literature review, data collection secondary structure of proteins, PSSM feature extraction, physicochemical feature extraction, SVM models, testing and evaluation. Flowchart of the research methods can be illustrated in Figure 1.

### *2.1. Literature review*

Things to do at this stage is to study the subjects related to protein secondary structure prediction and classification method of SVM. From the study, modeling with SVM classification can be implemented and used to predict the secondary structure of proteins by the addition of physicochemical features to improve the accuracy of the classification.

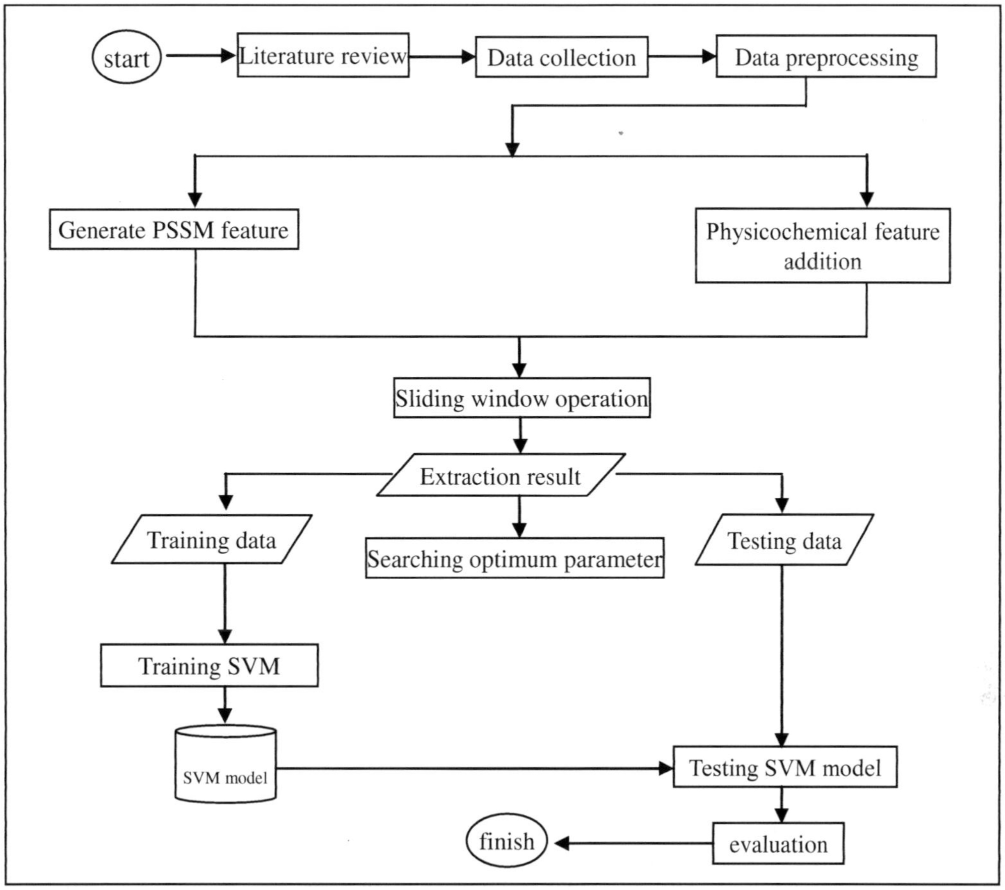


**Figure 1.** Flowchart of research methods

### *2.2. Data Collection and preprocessing*

The data retrieval process begins with collecting a secondary protein sequence data through the website of PDB in http://www.rcsb.org/ to get PDB id. The id indicating different structure We choose enzim sequence based on enzim commission information. Furthermore, the process generates a file DSSP (Dictionary Protein Secondary Structure). DSSP is database of protein secondary structure assignment. Tools used to generate file DSSP is XSSP in http://www.cmbi.ru.nl/xssp website. In this study, the existing class segment in the DSSP is reduced to three classes, namely alpha-helix (H), beta-sheet (B) and the coil (C) [13]. Detail of secondary structure assignment for data grouping can be shown in Table 1.

**Table 1**. Grouping of protein secondary data

| DSSP 8-Class | 3-Class |
|---|---|
| α-Helix (H) ,3/10 helix (G) | Helix (H) |
| β-Helix (E), β-Bridge (B) | Strand(E) |
| Π-Helix (I), Turn (T), Bend (S), Coil (C) | Coil (C) |

On preprocessing task, we just use two column in DSSP file, amino acid (aa) and structure. Amino acid describe the residue of sequence protein and structure inform us as secondary structure assignment.

### *2.3. Feature Extraction of PSSM*

In conducting protein secondary structure prediction, the features required to help define the secondary structure. The data from 20 amino acids used as a determinant of input pattern classification used. The data file that is processed in DSSP will be formed into PSSM profile. Forming PSSM profile is done by entering DSSP amino acids data files with tools on the web http://www.cbs.dtu.dk/biotools/Seq2Logo. To extract the features of PSSM profile, Sliding window used for pattern recognition process. The use of sliding window will acquire more information from protein residues with the optimal number of windows 13[5].

### *2.4. Physicochemical feature addition*

Physicochemical features of each amino acid were added to PSSM as input for the classification process. Physicochemical features will work as identifier in this study that includes conformation parameters, net charge, hydrophobic side chain mass[5]. The value of conformation parameters indicates the chances each amino acid residue of the secondary structure of H, E, and C. Net charge feature obtained by the index table of amino acids index. And hydrophobic side chain mass is used as a feature in the protein secondary structure prediction because it is associated with the folding process.

### *2.5. Sliding window operation*

Combination between PSSM and physicochemical are main focus on this research. The combination of two features then operated via sliding window operation to generate number of feature as input of Support Vector Machine. Output of this process as data model known Model I, model II and Model III that will be explained in next section.

### *2.6. Searching Optimal SVM parameter and training*

SVM as classifier fore predicting secondary structure must be tuned to obtain optimal parameter. Empirically, we use kernel on SVM from linear, radial and polynomial kernal with 2, 3 and 4 degree. Before training the model, one optimal kernel must be obtained first to ensure that error in training process can be minimized. The output of this process is optimal parameter for the kernel for training.

### *2.7. Classification method Support Vector Machine (SVM)*

In this study, dataset was built to three models in order to compare the accuracy of the results between the three models. The first model (Model I) was built with data that have been done PSSM feature extraction without the addition of physicochemical features do the sliding windows. The second model (Model II) was built with data that have been done and done PSSM feature extraction process of sliding windows. Then added six physicochemical features. The third model (Model III) was built with feature extraction data has been done PSSM then added six physicochemical features. Then do the process of sliding windows.

The difference between the three models is the number of features in each model into a dimension of a vector space SVM. Model I without addition of physical-chemical features while Model II and III are additional features with different physical chemistry physical chemistry feature placement position in data processing. The number of feature describe in greater detail below. Model I has 260 features obtained from 20 (amino acid) x 13 (sliding window). Model II with 266 features obtained from 260 + 6 (physicochemical). Last but not least, Model III position sliding window operation after physicochemical were added. Therefore, we obtain (20+6) x 13 = 338 features.

Before the classification process, the distribution of the data in the dataset is divided into two parts, the training data part as much as 75% of the dataset and test data part as much as 25% of the dataset. Distribution of the proportion of secondary structure classes can be seen in Figure 2.

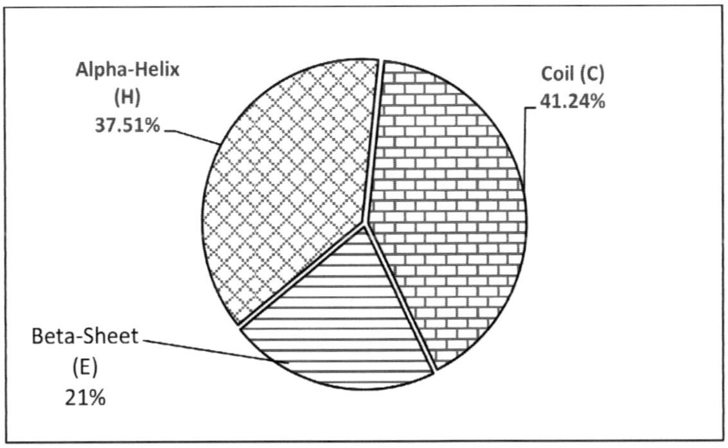


**Figure 2**. Data distribution on dataset

### *2.8. Testing and Evaluation*

Testing is performed to test the model generated by Support Vector Machine. Q3 accuracy assessment will use a score that contains information of data classes and predicted results. The value equation of accuracy using the equation 1 where i is the class type of protein secondary structure, alphahelix (H), beta-sheet (B), and coil (C). Equation 2 is the Q3 score equation which is the average accuracy value of the entire class. In Q3 similarity score, $N_H$ is accuracy for alphahelix (H), $N_B$ is accuracy to beta-sheet (B), $N_C$ is accuracy to coil (C), and $N_{total}$ is the number of classes on the test data.

$$Accuracy = \frac{\sum_{i=1}^{n} data\ of\ true\ i\ class}{\sum_{i=1}^{n} number\ of\ class\ i} \times 100\% \quad (1)$$

$$Q3\ Score = \frac{N_H + N_B + N_C}{N_{total}} \quad (2)$$

## 3. Result and Discussion

### *3.1. Data Collection and preprocessing*

At this stage, the protein sequence data retrieved from the file extension .dssp. Inside that file, there are 19 fields of data. This study only takes two columns, AA column (Amino Acid) and Structure. Secondary structures grade that exists on Structure column will be reduced from 8 class into 3 classes. Illustration of retrieval data column can be seen in Figure 3 and the reduction classes are conducted based on Table 1.

IOP Conf. Series: Journal of Physics: Conf. Series **835** (2017) 012006 doi:10.1088/1742-6596/835/1/012006

| # | Residue | AA | Structure | BP1 | BP2 | ... | Z-CA |
|---|---|---|---|---|---|---|---|
| 1 | 108B | I | H | 0 | 0 | | -36.2 |
| 2 | 109B | G | H | 0 | 0 | | -32.7 |
| 3 | 110B | S | T | 0 | 0 | ... | -34.2 |
| 4 | 111B | K | T | 0 | 0 | | -34.0 |
| 5 | 112B | G | S | 0 | 0 | | -32.4 |

| AA | Structure |
|---|---|
| I | H |
| G | H |
| S | T |
| K | T |
| G | S |

| AA | Structure |
|---|---|
| I | H |
| G | H |
| S | C |
| K | C |
| G | C |

**Figure 3**. Data collection illustration

### *3.2. Feature Extraction of PSSM*

After getting the protein sequence data collection, the sequence included as input in http://www.cbs.dtu.dk/biotools/Seq2Logo/ to do the generation characteristic of amino acids process. This process is a search for value changes in amino acid residues of the other amino acid residues. Interesting patterns contained in the value of the change is an amino acid value becomes highest when met with the amino acid itself. Once the protein sequence is inserted, then the resulting PSSM matrix measuring the number of inputs × 20 (amino acids). PSSM matrix form can be seen in Figure 4.

```
A R N D C Q E G H I L K M F P S T W Y V
1 I -1.362 -3.086 -3.248 -3.166 -0.136 -3.195 -3.672 -3.465 -2.697 4.483 1.691 -2.664 1.286 0.232 -3.49 -3.062 -1.06 -2.17 -1.271 2.875
2 G 0.104 -2.343 -0.426 -1.308 -1.306 -2.181 -2.608 5.796 -1.462 -3.201 -3.446 -1.636 -2.652 -2.664 -2.765 -0.988 -1.905 -2.426 -3.134 -2.85
3 S 1.082 -0.727 0.511 -0.234 0.087 -0.555 -0.541 -0.092 -0.446 -1.897 -2.314 -0.271 -1.172 -1.929 -1.461 3.212 1.025 -2.535 -1.741 -1.278
4 K -0.812 2.108 -0.254 -0.708 -1.963 0.821 0.328 -1.328 -0.229 -2.109 -2.226 4.453 -1.209 -2.791 -1.673 -0.867 -1.068 -2.535 -1.774 -1.984
5 G 0.104 -2.343 -0.426 -1.308 -1.306 -2.181 -2.608 5.796 -1.462 -3.201 -3.446 -1.636 -2.652 -2.664 -2.765 -0.988 -1.905 -2.426 -3.134 -2.85
```

**Figure 4**. Matrix PSSM Illustration

### *3.3. Physicochemical feature addition*

At this stage of the calculation of the parameter value parameter conformation, net charge, hydrophobic, and side chain mass. Conformation parameter is the composition of the amino acid sequence of a particular class of secondary structures that exist on the dataset. Net charges are physical-chemical parameters which only five amino acid arginine (R), aspartic acid (D), glutamine acid (E), histidine (H), and lysine (K). This relates to the formations that will be formed by the electric charge on the five amino acids. Values electrical charge provided are +1 and -1 selainnya is worth 0. Hydrophobic properties that affect the stability of the secondary structure of proteins. The more positive the value of these properties, the hydrophobic nature of the stronger. Side chain mass is a value that is calculated based on the mass of the chemical bond, an amino acid that affects protein structure folding process. In this study, the parameter value parameter conformation, net charge, hydrophobic, and side chain mass.

### *3.4. Sliding window operation*

This method is a process of feature extraction from PSSM matrix with secondary structure proteins class column written derived from tables of data retrieval. This process uses sliding windows which amounted to 13 refer to [5]. Windows that exist in the sliding window will take the data or protein sequence as many as 13 rows with row-7 as a point of interest and secondary structure are assigned as class label to this point of interest. The result of this process becomes the input data or the dataset used for the purposes of training and testing process. There are three data model are generated from this process. Data with 260, 266 and 338 respectively. Number of features are described on subsection 2.7

### *3.5. SVM Optimum Parameters*

Empirically, combination of kernel and parameter are examined at this stage. We spent eight hours from our research for getting optimal parameter. C and $\gamma$ are parameter will be tuned. The parameter C and $\gamma$ used in the classification of SVM with radial and polynomial kernel. The data used were 9000 data from the dataset and tuning process is carried out with a grid search method. The function used is tune.svm() from e1071 R package. The parameters C and $\gamma$ are determined by a combination of $5 \times 4$ matrix. The combination matrix formed from $[2^{-2}, 2^{-1}, 2^{0}, 2^{1}, 2^{2}]$ and $[2^{-6}, 2^{-5}, 2^{-4}, 2^{-3}]$. Optimum parameters obtained by the $C = 2^1$ and $\gamma = 2^{-6}$.

### *3.6. Support Vector Machine Classification*

At this stage performed the optimum parameter search, SVM model training, and testing of SVM models. Searches conducted to find optimum parameters for training with radial and polynomial kernel. Package used in this study is a package e1071 with svm(), predict(), and tune.svm() function. The details of the process are as follows:

### *3.7. Training Model SVM*

At this stage, SVM model training conducted with training data which has been divided from the dataset with the optimum parameters of the search parameters. Training is done on every model with linear kernel, kernel radial and polynomial kernel of degree 2, 3, and 4. The resulting model as many as 15 pieces. Training SVM models using svm() function from e1071 R package.

### *3.8. Testing Model SVM*

In this stage, testing conducted to models that have been trained with a dataset and a different kernel. Tests conducted by the predict() from e1071 R package. The test results in the chart of Q3 Score values can be illustrated in Figure 5.

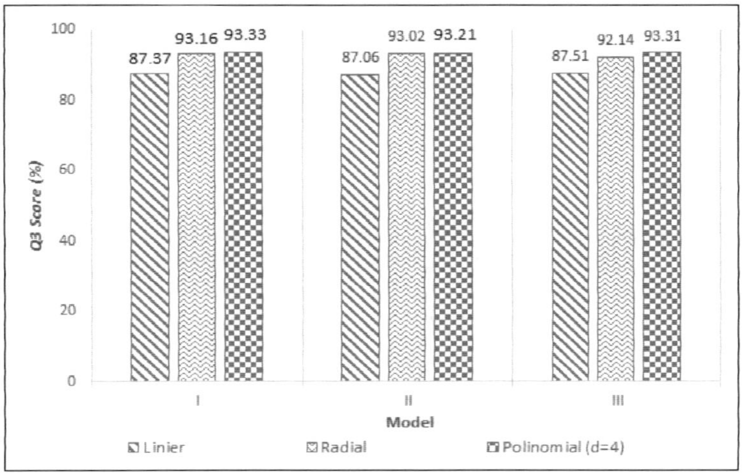


**Figure 5**. Q3 Score diagram

The figure 5 indicate that the highest score value Q3 owned by model I with polynomial kernel of degree 4. Model I have 260 PSSM profile features without physical chemical features. Values obtained is 93.33%. Lowest score Q3 Value owned by Model II with linear kernel. Model II has 266 features (260 features six feature PSSM profile and physicochemical). Values obtained for this model is 87.06%. The highest Q3 value score obtained by the model dataset is processed by the polynomial kernel of degree 4. From the figure 5 indicate that the addition of physicochemical after and before sliding window not have significance accuracy on prediction. This is because the values of physicochemical are static and relatively too little compared to PSSM features so does not give more information as a feature.

## 4. Conclusion

Secondary structure can be performed using SVM with PSSM and physicochemical as feature. Furthermore, evaluation of models produced by calculating the value of Q3 Score. Additional features of physical chemistry can be implemented, however does not have a significant influence on the value of accuracy. The best results obtained by the model derived from the dataset Model I by using kernel polynomial of degree four.

## References

[1] Polanski A and Kimmel M 2007 *Bioinformatics* (New York: Springer)
[2] Bray A, Johnson H, Raff L and Walter R 2009 *Essential cell biology* Ed 3th (London: Garland Science)
[3] Zhuravleva A and Korzhnev D M 2017 *Prog. Nucl. Magn. Reson. Spectrosc.* **100** 52–77
[4] Martin J, Letellier G, Marin A and De Brevern A G 2005 *BMC Structural Biology* **17** 1–17
[5] Huang Y and Chen S 2013 *Sci. World J.* **2013** 1–8
[6] Wang J, Wang C, Cao J, Liu X, Yao Y and Dai Q 2015 *Gene* **554** 241–8
[7] Vapnik V N and Cortes C 1995 *Machine Learning* **20** 273–97
[8] Patel M S and Mazumdar H S 2014 *J. Theor. Biol.* **361** 182–9
[9] Atar E, Ersoy O, and Ozyilmaz L 2005 *2005 ICSC Congress* (Istanbul: Computational Intelligence Methods and Applications)art. no. 1662358
[10] Drozdetskiy A, Cole C, Procter J and Barton G J 2015 *Nucleic Acids Res* **43** 389–94
[11] Yang B, Wu Q, Ying Z and Sui H 2011 *Knowledge-Based Systems* **24** 304–13
[12] Johal A K 2014 *International Journal Of Engineering And Computer Science* **3** 3593–7
[13] Aydin Z, Altunbasak Y and Borodovsky M 2006 *BMC Bioinformatics* **7** 178

IOP Conf. Series: Journal of Physics: Conf. Series **835** (2017) 012007 doi:10.1088/1742-6596/835/1/012007

# Metagenomics workflow analysis of endophytic bacteria from oil palm fruits

**Z A Tanjung*, R Aditama, W M Sudania, C Utomo, and T Liwang**
Plant Production and Biotechnology Division, PT SMART Tbk.,
Sinar Mas Land Plaza, 2nd Tower, 10th Fl., Jl. M.H. Thamrin No. 51, Jakarta 10350, Indonesia

E-mail: biotechnology@sinarmas-agri.com

**Abstract.** Next-Generation Sequencing (NGS) has become a powerful sequencing tool for microbial study especially to lead the establishment of the field area of metagenomics. This study described a workflow to analyze metagenomics data of a Sequence Read Archive (SRA) file under accession ERP004286 deposited by University of Sao Paulo. It was a direct sequencing data generated by 454 pyrosequencing platform originated from oil palm fruits endophytic bacteria which were cultured using oil-palm enriched medium. This workflow used SortMeRNA to split ribosomal reads sequence, Newbler (GS Assembler and GS Mapper) to assemble and map reads into genome reference, BLAST package to identify and annotate contigs sequence, and QualiMap for statistical analysis. Eight bacterial species were identified in this study. *Enterobacter cloacae* was the most abundant species followed by *Citrobacter koseri, Seratia marcescens, Latococcus lactis* subsp. *lactis, Klebsiella pneumoniae, Citrobacter amalonaticus, Achromobacter xylosoxidans,* and *Pseudomonas* sp. respectively. All of these species have been reported as endophyte bacteria in various plant species and each has potential as plant growth promoting bacteria or another application in agricultural industries..

## 1. Introduction

Next-Generation Sequencing (NGS) or high-throughput sequencing has become a powerful sequencing tool for microbial study. It has led to the establishment of the field of "meteganomics" [1]. Metagenomics (also referred to as environmental and community genomics) is the microorganisms genomic analysis of microorganisms by direct extraction and cloning of DNA from an assemblage of microorganisms [2]. Generally, it is used to investigate marines [3] and soils [4] complex microbial communities. NGS consists of different platforms, such as Roche 454 pyrosequencing and Illumina/Solexa as a leading platforms. Roche 454 pyrosequencing has increase its average read lengths to ~500 base pairs compared to Illumina which only ~150 base pairs [5].

Endophytic bacteria have wide range of effects on their hosts, such as: increasing phosphate solubilization and nitrogen fixation, repressions of soil-borne pathogens, breakdown of plant-produced ethylenes by bacterial production of 1-aminocyclopropane-1-carboxilate deaminase, induced systemic

* Corresponding author

resistances, and synthesis of phytohormones [6]. Oil palm fruits are the most efficient source of vegetable oil. An endophyitic bacteria that may live in oil palm fruits certainly have produced enzyme activity which could utilize vegetable oil for industrial application. This study described a workflow to analyze metagenomics data via direct sequencing from endophytics bacteria in oil palm fruits.

## 2. Material and Methods

### *2.1 Materials*

The raw data of this study was Sequence Read Archive (SRA) file under accession ERP004286. This SRA was uploaded by University of Sao Paulo. SILVAdb was used for identification of RNA sequences [7]. SWISS-PROT database [8] and NCBI-NR database were used for functional annotations.

The bioinformatics software used for this study were SRA Toolkit [9], Filter FASTQ [10], SortMeRNA [11], Newbler (GS Assembler and GS Mapper) [12], BLAST package [13], and QualiMap [14].

### *2.2 Methods*

All methods used in this study were described in Figure 1. This workflow was designed for analyzing direct sequencing data from metagenomics analysis of oil palm fruits using 454 pyrosequencing platform.

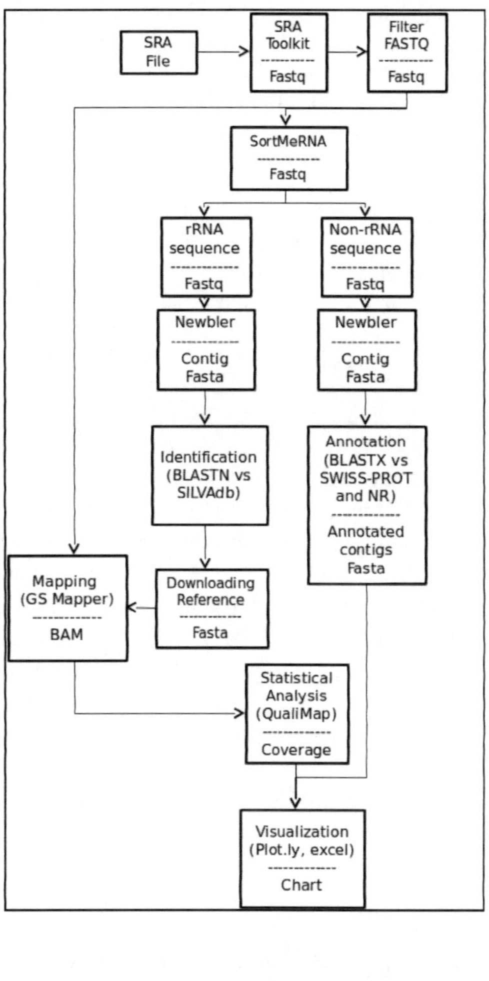


**Figure 1.** Metagenomics workflow analysis of endophytic bacteria from oil palm fruits

## 3. Results and Discussion

This study used Sequence Read Archive (SRA) file under accession ERP004286 that contained raw data of metagenomics sequencings of endophytic bacteria in oil palm fruits. The bacterial endophytes within the oil palm fruits (*Elaeis guineensis*) were cultured using an oil-palm enriched medium and the bacterial cells present in the microcosm were separated by filtration through a 14 micrometer pore filter. The DNA obtained by these cells was then sequenced using 454 pyrosequencing technology on the 454 GS FLX platform. A single fragment (shotgun) library was sequenced in two separate runs. It was generated from 454 pyrosequencing platforms and has single-end reads.

The raw data have 548,757 reads. After trimming process, it was assembled by GS Assembler (Newbler). It produced 74 contigs from ribosomal sequence and 18,600 contigs from non-ribosomal sequence. The ribosomal sequence was identified using blastn against SILVAdb. This identified 8 species which live in oil palm fruits. The microbial diversity profile of endophytic bacteria in oil palm fruits was described on Figure 2. There were 7 species which belong to Phylum Proteobacteria and 1 species belong to Pyhlum Firmicutes. *Lactococcus lactis* subsp. *lactis* was the only species belong to Firmicutes. It was known as plant-associated lactic acid bacterium. *Enterobacter cloacae* was the most abundant in oil palm fruits. It is followed by another Enterobacteriaceae, which were *Citrobacter koseri, Seratia marescens, Klebsiella pneumoniae, Citrobacter amalonaticus,* and *Pseudomonas* sp.. *E. cloacae* was known as plant growth-promoting bacteria (PGPB). It was also an endophytic symbiont of corn, cacao, and pines [15, 16, 17]. *C. koseri* was also reported as endophyte in corn and cotton [18]. *S. marcescens* has been reported as cold tolerance and plant growth-promotion potential from summer squash [19]. *K. pneumoniae* was reported as nitrogen-fixing bacteria in wheat [20]. *C. amalonaticus* have phytase gene which potentially to degrade phytic acid [21]. *A. xylosoxidans* was the only bacteria member of Alcaligenaceae. It was reported as plant growth promoting bacterium in wheat [22] and as a copper-resistant plant growth promoting bacteria in mustard [23].

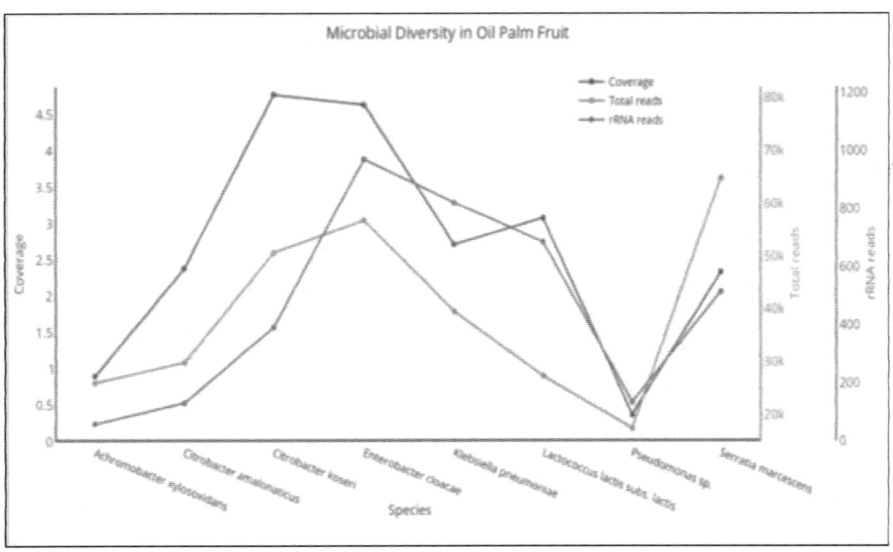


**Figure 2.** Microbial diversity profile from endophytic bacteria of oil palm fruits

There are limited publication of metagenomics profile of bacteria in plant fruits. It was done in strawberry and rambutan fruits by selected-colony to 16S sequencing [24,25] also in citrus by PCR-DGGE [26]. Bacterial profile in strawberry is more likely to oil palm fruits than citrus and rambutan fruits. Genus *Enterobacter, Lactobacillus,* and *Pseudomonas* also present in strawberry fruits [24].

Metagenomics in plant fruits analysis using whole genome shotgun sequencing give the real abundance of bacterial community profile because it reduces PCR bias compare to 16S-NGS method. In this study, 454 pyrosequencing platform was used. It give medium long reads around 600-700 bp compare to Illumina platform which generate 250 bp at most. This study provide a workflow to analyse whole genome shotgun metagenomics 454 pyrosequencing platform data of oil palm fruits to produce bacterial community profile.

## 4. Conclusion

This workflow performed well in revealing ribosomal and non-ribosomal metagenomics sequences of endophytic bacteria from oil palm fruits. According to their coverage, *E. cloacae* is the most abundant and followed by *C. koseri, S. marcescens, L. lactis* subsp. *lactis*, *K. pneumoniae*, *C. amalonaticus*, *A. xylosoxidans* and *Pseudomonas* sp.. Each species has their own potential in agricultural applications.

## References

[1] Oulas A, Pavloudi C, Polymenakou P, Pavlopoulos GA, Papanikolaou N, Kotoulas G and Iliopoulos I 2015 *Bioinform Biol Insights* **9** 75–88
[2] Handelsman J 2004 *Microbiol Mol Biol Rev*. **68** 669–85
[3] Gilbert J, Field D, Huang Y, Edwards R, Li W, Gilna P and Joint I 2008 *PloS one. 3* e3042
[4] Urich T, Lanzén A, Qi J, Huson DH, Schleper C and Schuster SC 2008 *PloS one. 3* e2527
[5] Balzer S, Malde K, Lanzén A, Sharma A and Jonassen I 2010 *Bioinformatics* **26** i420–5
[6] De Melo Pereira GV, Magalhães KT, Lorenzetii ER, Souza TP and Schwan RF 2012 *Microb. Ecol.* **63(2)** 405–17
[7] Pruesse E, Quast C, Knittel K, Fuchs BM, Ludwig W, Peplies J and Glöckner FO 2007 *Nucleic Acids Res.35* 7188–96
[8] Consortium TU 2014 *Nucleic Acids Res.* **43** 204–12
[9] Leinonen R, Sugawara H and Shumway M 2011 *Nucleic Acids Res.* **39** D19–21
[10] Blankenberg D, Gordon, Von Kuster G, Coraor N, Taylor J and Nekrutenko 2010 *Bioinformatics* **26** 1783–5
[11] Kopylova E, Noé L and Touzet H 2012 *Bioinformatics* **28** 3211-7
[12] Margulies M, Egholm M, Altman WE, Attiya S, Bader JS, Bemben LA and Rothberg JM 2005. *Nature* **437** 376–80
[13] Altschul SF, Gish W, Miller W, Myers EW and Lipman DJ. 1990. *J. Mol. Biol*. **215** 403-10.
[14] García-Alcalde F, Okonechnikov K, Carbonell J, Cruz LM, Götz S, Tarazona S, and Conesa A 2012 *Bioinformatics* **28** 2678–9
[15] Hinton DM, Bacon CW and Ars U 1995 *Mycopathologia* **129** 117–25
[16] Zayda PMM, Elizabeth AAD, Thiago ASDO, Fernando PM, Leandro LL and Jorge TDS 2015 *Afr. J. Microbiol. Res.* **9** 1352–56
[17] Liu WY, Chung KMK, Wong CF, Jiang JW, Hui, RKH and Leung FCC 2012 *J. Bacteriol.* **194** 5965
[18] McInroy JA and Klepper JW 1995 *Plant Soil* **173** 337-42
[19] Selvakumar G, Mohan M, Kundu S, Gupta AD, Joshi P, Nazim S and Gupta HS 2008 *Lett. Appl. Microbiol.* **46** 171–5
[20] Iniguez AL, Dong Y and Triplett EW 2004 *Mol. Plant. Microbe. Interact.* **17** 1078–85

[21] Li C, Lin Y, Huang Y, Liu X and Liang S 2014 *PloS One* **9** e114728
[22] Jha P and Kumar A 2009 *Microb. Ecol.* **58** 179–88
[23] Ma Y, Rajkumar M and Freitas H 2009 *J. Environ. Manage.* **90** 831–7.
[24] Pereira GVDM, Magalhães KT, Lorenzetii ER, Souza TP and Schwan RF 2012 *Microb Ecol.* **63** 405–17
[25] Suhandono S, Kusumawardhani MK and Aditiawati P 2016 *HAYATI J. Biosci.* **23** 39–44
[26] Nguyen DDL, Gemrot E, Loiseau G and Montet D 2008 *Fruits* **63** 75–84

**Acknowledgement**
The authors gave utmost gratitude to the management of PT SMART Tbk, for supporting the publication of this research paper.

IOP Conf. Series: Journal of Physics: Conf. Series **835** (2017) 012008 doi:10.1088/1742-6596/835/1/012008

# Vitual screening and binding mode elucidation of curcumin analogues on Cyclooxygenase-2 using AYO_COX2_V1.1 protocol

**E. Mulatsari***, **E. Mumpuni and A. Herfian**

Department of Pharmacy, Pancasila University, Jakarta,12640, farmasi@ffup.org, Indonesia

E-mail: estimulatsari@gmail.com

**Abstract**. Curcumin is yellow colored phenolic compounds contained in Curcuma longa. Curcumin is known to have biological activities as anti-inflammatory, antiviral, antioxidant, and anti-infective agent [1]. Synthesis of curcumin analogue compounds has been done and some of them had biological activity like curcumin. In this research, the virtual screening of curcumin analogue compounds has been conducted. The purpose of this research was to determine the activity of these compounds as selective Cyclooxygenase-2inhibitors in in-silico. Binding mode elucidation was made by active and inactive representative compounds to see the interaction of the amino acids in the binding site of the compounds. This research used AYO_COX2_V.1.1, a structure-based virtual screening protocol (SBVS) that has been validated by Mumpuni E et al, 2014 [2]. AYO_COX2_V.1.1 protocol using a variety of integrated applications such as SPORES, PLANTS, BKchem, OpenBabel and PyMOL. The results of virtual screening conducted on 49 curcumin analogue compounds obtained 8 compounds with 4 active amino acid residues (GLY340, ILE503, PHE343, and PHE367) that were considered active as COX-2 inhibitor.

## 1. Introduction

Curcumin is known to have biological activities as anti-inflammatory, antiviral, antioxidant, and anti-infective agents [1]. However, curcumin is an unstable compound, because it is easily degraded through hydrolysis at neutral and alkaline pH [3]. Therefore, A range of research has been done to modify the curcumin compound such as works conducted by Nurfina et al [4], Sardjiman et al [5] and Supardjan et al [6]. Several of synthesized curcumin analogues compounds, through the in vivo and in vitro test, showed biological activity as a curcumin. One of the biological activities is the ability to inhibit the cyclooxygenase-2 enzyme (COX-2).

COX-2 enzyme is responsible for inflammatory and cancer mediator. Overexpression of COX-2 in cancer cells resulting in overproduction of prostanoid leading to increase of cell proliferation and prevention of apoptosis. In colon cancer cells, the overproduction of the COX-2 enzyme is also accelerated by the process of angiogenesis. It is caused by the catalysis product of COX-2 that will stimulate the activity of angiogenic factors [7, 8]. This shows the importance of research to find compounds that can inhibit the action of the COX-2 enzyme as the design of drug discovery. Utilization of computers in the study of drug discovery, *in silico* methode, is an analogue with *in vitro*

* Corresponding author

and *in vivo*. The approach of virtual screening was used to identify active compounds based on structure of certain enzymes, so called the Structure-Based Virtual Screening (SBVS), which simulated docking molecules often used as a basis for the development of SBVS protocols. Molecular docking, used in the study SBVS is regarded as an efficient way in drug discovery. SBVS protocol in this study is the protocol built for identifying COX-2 inhibitors compounds.

The protocol is called AYO_COX2_V.1.1 that has been validated by E. Mumpuni. et al, 2014 [2]. AYO_COX2_V.1.1 protocol is validated retrospectively to see the value of Enrichment Factor (Comparative between the percent of True Positive and False Positive percent.) as the objective function by using the default settings and modified settings. In the default setting obtained value of EF20% = 3.016 and the value of EFmax = infinity (~), had the quality of structure-based screening protocol as "excellent" by reference [9]. In addition to using EF20% and EFmax, the protocol used an assessment of the value of Matthews correlation coefficient (MCC) and the EF value at the time the highest MCC value (MCCmax). The MCCmax value was used to find ligands which were used as comparison ligand that capable to distinguish active and inactive compounds for virtual screening. MCCmax value = 0.188 and EF values when MCCmax = 4,105 which was the value of active ligand on the DUD data set with the code ZINC03814604. The ligand can be used as comparison ligand. Furthermore, This study used a protocol that has been validated namely the default protocol with reference compounds ZINC03814604 to find COX-2 inhibitors. The results are expected to obtain curcumin analogues compounds which have activity as selective COX-2 inhibitors in *in silico*, so that it can be analyzed as more potential of curcumin analogues compounds that can be used as candidate drugs known as coxib (COX-2 inhibitor).

## 2. Methodology

### 2.1. Construction of the test compounds

In terminal mode applications were used AYO_COX2_V.1.1 with command line interface with the command: ./AYO_COX2_v.1.1.sh. *BkChem* application window appeared and each test compounds in Table 2.1 was drawn in the form of two-dimensional and exported with the file name uji.mol. After that, the AYO_COX2_V.1.1 application returned ChemPLP score that stored with the file name hasil.txt, besides the scores of test compounds on the results file also found ChemPLP score of comparison compounds, this steps were done by replication at least 3 times. Binding mode elucidation has done by selected representative active and inactive compounds were used for visualizing using *PyMOL* to do elucidation modes of bonding in the receptor binding site and determining the amino acid residues that actively interacted with the ligand as a COX-2 inhibitor.Positive control test was also conducted using the compound of the drugs known as coxibs.

## 3. Results and Discussion

Virtual screening was done by preparing the test compounds and comparison compounds. Each compound was drawn using *BKchem* applications that opened automatically when running SBVS protocol. Screening was done at least three times replication to obtain the docking simulation results score called *chemPLP* score which was the total energy calculated from one docking process [9]. *ChemPLP* Score of the test compounds was compared with the comparison compound ZINC03814604. The test compounds were active in *in silico* on COX-2 binding pocket if it had a smaller score (more negative) than the score of comparison compound, meaning that the compound had good stability. The docking simulation results and in *in silico* compound activity are presented in Table 1.

Based ChemPLP score obtained 8 predicted active compounds i.e : 1,7-Diphenil-1,6-heptadien-3,5-dione, 1,7-bis(4-hydroxyphenyl)-1,6-heptadien-3,5-dione, 1,7-bis(3-methoxyphenyl)-1,6-heptadien-3,5-dione, 1,7-bis(4-hydroxy-3,5-dimethylphenyl)-1.6-heptadien-3,5-dion, 1,7-bis(3,5-diethyl-4-hydroxyphenyl)-1.6-heptadien-3,5-dione, 1,7-bis(2-methoxyphenyl)-1.6-heptadien-3,5-dione, 1,7-bis(4-chlorophenyl)-1.6-heptadien-3,5-dione, 1,7bis (4-methylphenyl) -1.6-heptadien-3,5-dione, with the average of ChemPLP score respectively : -91.9147, -86.1098, -90.2596, -88.5337, -90.5843, -95.8059, -87.3040, and -88.0893. Based on the *ChemPLP* score, 1,7-bis (2-methoxyphenyl) -1.6-heptadien-3,5-dione had the highest affinity in the receptor COX-2 because it had the smallest *ChemPLP* score (most negative) than the another test compound.

Having obtained a score of each test compound and the comparison compound, then statistical analysis of one-tailed paired t-test was performed. In statistical testing, if p-value was less than 0.05 at the 95% confidence level, it can be stated that average score of test compound was significant or meaningful that can represent the entire population. The predicted active compounds as COX-2 inhibitors are 1,7-Diphenil 1,6-heptadien-3,5-dione, 1,7-bis (4-hydroxyphenyl) -1.6-heptadien-3, 5-dione, 1,7-bis (3-methoxyphenyl) -1.6-heptadien-3,5-dione, 1,7-bis (4-hydroxy-3,5-dimethylphenyl) -1.6-heptadien- 3,5-dione, 1,7-bis (3,5-diethyl-4-hydroxyphenyl) -1.6-heptadien-3,5-dione, 1,7-bis (2-methoxyphenyl) -1,6- heptadien-3,5-dione, 1,7-bis (4-chlorophenyl) -1.6-heptadien-3,5-dione, 1,7-bis (4-methylphenyl) -1.6-heptadien-3, 5-dione with a p-value respectively 0.0002, 0.0087, 0.0013, 0.0011, 0.0008, 0.0002, 0.0085, 0.0004. All of them had a p-value less than 0.05, this showed that all of these compounds had activity as active inhibitors of COX-2.

**Table 1.**ChemPLP score and inhibitors activity of test compounds and comparison compound

| Test Compounds | ChemPLP score of test compounds | ChemPLP score of comparison compound | COX-2 inhibitor Activity |
|---|---|---|---|
| **Natural analogue curcumin as reference compound** | | | |
| Curcumin (Enol) | -86.2946 | -84.7546 | active |
| Curcumin (Keto) | -87.4828 | -84.6900 | active |
| Desmethoxycurcumin | -85.5663 | -84.7107 | active |
| Bidesmethoxycurcumin | -86.1965 | -84.7078 | active |
| **Analogue Curcumin CompoundsSynthesized by Nurfina et al 1997 [4]** | | | |
| 1,7-Diphenyl-1,6-heptadien-3,5-dion | -91.9147 | -84.7398 | **active** |
| 1,7-bis(4-hydroxyphenyl)-1,6-heptadien-3,5-dion | -86.1098 | -84.5319 | **active** |
| 1,7-bis(4-hydroxy-3-methoxyphenyl)-1,6-heptadien-3,5-dion | -82.5414 | -84.7290 | inactive |
| 1,7-bis(3-methoxyphenyl)-1,6-heptadien-3,5-dion | -90.2596 | -84.7467 | **active** |
| 1,7-bis(3,4,5-trimethoxyphenyl)-1,6-heptadien-3,5-dion | -70.7774 | -84.7197 | inactive |
| 1,7-bis(4-hydroxy-3,5-dimethylphenyl)-1,6-heptadien-3,5-dion | -88.5337 | -84.6948 | **active** |
| 1,7-bis(3,5-diethyll-4-hydroxyphenyl)-1,6-heptadien-3,5-dion | -90.5843 | -84.7361 | **active** |
| 1,7-bis(4-hydroxy-3,5-diisopropylphenyl)-1,6-heptadien-3,5-dion | -78.2153 | -84.7174 | inactive |
| 1,7-bis(2-methoxyphenyl)-1,6-heptadien-3,5-dion | -95.8059 | -84.7337 | **active** |
| 1,7-bis(4-chlorophenyl)-1,6-heptadien-3,5-dion | -87.3040 | -84.7251 | **active** |
| 1,7-bis(4-methoxyphenyl)-1,6-heptadien-3,5-dion | -84.1965 | -84.7284 | inactive |
| 1,7-bis(4-methylphenyl)-1,6-heptadien-3,5-dion | -88.0893 | -84.7441 | **active** |
| 1,7-bis(3,4-dimethoxyphenyl)-1,6-heptadien-3,5-dion | -81.4540 | -84.7291 | inactive |
| 1,7-bis(3,5-Di-tert-buthyl-4-hydroxyphenyl)-1,6-heptadien-3,5-dion | -37.2678 | -84.7282 | inactive |
| 1,7-bis(3,4-dibenzylioksphenyl)-1,6-heptadien-3,5-dion | -33.4720 | -84.7070 | inactive |
| 1,7-bis(4-hydroxy-3,5-dimethoxyphenyl)-1,6-heptadien-3,5-dion | -75.4518 | -84.7206 | inactive |

**Table 1.** (Continue)

| | | | |
|---|---|---|---|
| **Analogue Curcumin CompoundsSynthesized by Sardjiman et al 1997 [5]** | | | |
| 2,6-bis(4-hidroxybenzyl)cyclohexanone | -68.8778 | -84.7441 | inactive |
| 2,6-bis(4-hydroxy-3-methoxybenzyl)cyclohexanone | -72.8071 | -84.7243 | inactive |
| 2,6-bis(4-hydroxy-3,5-dimethylbenzyl)cyclohexanone | -69.0245 | -84.7297 | inactive |
| 2,6-bis(3,5-diethyl-4-hydroxybenzyl)cyclohexanone | -71.1657 | -84.7287 | inactive |
| 2,6-bis(4-hydroxy-3,5-diisopropylbenzyl)cyclohexanone | -60.1065 | -84.7187 | inactive |
| 2,6-bis(3,5-di-tert-butyl-4-hydroxybenzyl)cyclohexanone | -22.7815 | -84.7077 | inactive |
| 2,6-bis(4-hydroxy-3,5-dimethoxybenzyl)cyclohexanone | -64.6337 | -84.7222 | inactive |
| 2,6-bis(3,5-dichloro-4-hydroxybenzyl)cyclohexanone | -69.9008 | -84.7135 | inactive |
| 2,5-bis(4-hydroxybenzyl)cyclopentanone | -77.0433 | -84.7305 | inactive |
| 2,5-bis(4-hydroxy-3-methoxybenzyl)cyclopentanone | -72.5252 | -84.7192 | inactive |
| 2,5-bis(4-hydroxy-3,5-dimethylbenzyl)cyclopentanone | -84.4206 | -84.7253 | inactive |
| 2,5-bis(3,5-diethyll-4-hydroxybenzyl)cyclopentanone | -84.0432 | -84.6667 | inactive |
| 2,5-bis(4-hydroxy-3,5-diisopropylbenzene)cyclopentanone | -72.8438 | -84.7381 | inactive |
| 2,5-bis(3,5-di-tert-butyl-4-hydroxybenzyl)cyclopentanone | -33.4951 | -84.7132 | inactive |
| 2,5-bis(4-hydroxy-3,5-dimethoxybenzyl)cyclopentanone | -72.4623 | -84.7088 | inactive |
| 2,5-bis(3,5-dichloro-4-hydroxybenzyl)cyclopentanone | -79.8052 | -84.7561 | inactive |
| 1.5-bis(4-hydroxyphenyl)1.4-pentadien-3-on | -76.6006 | -84.6985 | inactive |
| 1.5-bis(4-hydroxy-3-methoxyphenyl)1.4-pentadien-3-on | -76.0701 | -84.7191 | inactive |
| 1.5-bis(4-hydroxy-3.5-dimethylphenyl)1.4-pentadien-3-on | -78.469 | -84.7457 | inactive |
| 1.5-bis(4-hydroxy-3.5-dimethoxyphenyl)1.4-pentadien-3-on | -69.9995 | -84.7023 | inactive |
| 1.5-bis(3.5-dichloro-4-hydroxyphenyl)1.4-pentadien-3-on | -74.8726 | -84.7396 | inactive |
| **Analogue Curcumin CompoundsSynthesizedby Supardjan *et al* 1999[6]** | | | |
| 4-methyl-curcumin | -69.3172 | -84.7261 | inactive |
| 4-ethyllcurcumin | -71.5290 | -84.7385 | inactive |
| 4-n-propylcurcumin | -70.7039 | -84.7020 | inactive |
| 4-isopropylcurcumin | -67.5033 | -84.7318 | inactive |
| 4-n-buthylcurcumin | -72.1470 | -84.7258 | inactive |
| 4-benzylcurcumin | -55.4535 | -84.7254 | inactive |
| 4-phenylcurcumin | -56.2741 | -84.6900 | inactive |
| 4-(p-fluorophenyl)curcumin | -76.8823 | -84.7336 | inactive |
| 4-(p-methoxyphenyl) curcumin | -57.5191 | -84.7146 | inactive |
| 4-(p-methylphenyl )curcumin | -59.1340 | -84.7254 | inactive |
| 4-(m-trifluoromethylphenyl) curcumin | -60.4516 | -84.6687 | inactive |
| 4-(o,p-dinitrophenyl) curcumin | -44.3679 | -84.7177 | inactive |
| **Compounds of Coxib group as reference compounds** | | | |
| Celecoxib | -88.5473 | -84.7317 | active |
| Etoricoxib | -85.2023 | -84.7160 | active |
| 1.5-bis(3-etoksi-4-hydroxyphenyl)1.4-pentadien-3-on (Deetoksi EHP) | -85.2539 | -84.7298 | active |

*3.1.* Binding Mode Elucidation of Representative Compounds

Binding mode elucidation was depicted with *PyMOL* application. In the processes, desmethoxycurcumin, bisdesmethoxycurcumin, 1,7-bis (4-hydroxyphenyl) -1.6-heptadien-3,5-dion, 1,7-bis (3-methoxyphenyl) -1.6-heptadien-3,5-dione, 1,7-bis (4-hydroxy-3,5-dimethylphenyl)-1.6-heptadien-3,5-dion, 1,7-bis(3,5-diethyl-4-hydroxyphenyl)-1.6-heptadien-3,5-dione, 1,7-bis (2-methoxyphenyl) -1.6-heptadien-3.5 -dion, 1,7-bis (4-chlorophenyl) -1.6-heptadien-3,5-dione, 1,7-bis (4-methylphenyl) -1.6-heptadien-3,5-dion were used to identify amino acid residues that actively interacted with the active representative compounds. The number of amino acid residues that interact with all of them respectively 28, 20, 22, 24, 26, 26, 26, 24 and 25 amino acid residues. Binding mode elucidation of inactive representative compounds was done on 1,7-bis (3,4-dibenzyloksphenyl)-1.6-heptadien-3,5-dione and obtained 37 amino acids residues interaction. Amino acid residues that interacted with active and inactive representative compounds are shown in Table 2.

**Table 2.** Amino Acid Resiudes of Active and Inactive Representative Compounds

| Group | Representatif Compounds | Amino Acid Residues |
|---|---|---|
| Natural analogues curcumin | Demethoxycurcumin (active) | ARG106, ARG499,ALA513, GLY340, GLY512, GLY519, ILE98, LEU78, LEU338, LEU345,LEU370,LEU511, LEU517, MET508, PHE367, PHE504, SER339, SER516, TYR101, TYR334,TYR341, TYR371, TRP373, VAL74, VAL102, VAL330, VAL335, VAL509 |
| | Bidemethoxycurcumi n (active) | ARG106,ALA513,GLY512, GLY519, HIS75, ILE503, LEU338, LEU345, LEU511, LEU517, MET99, PHE504, SER339, SER516, TYR334,TYR341, TRP373, VAL330, VAL335, VAL509 |
| Derivatife compounds of 1,7-Diphenyl-1,6-heptadien-3,5-dione | 1,7-bis(3-methoxyphenyl)-1,6-heptadien-3,5-dione (active) | ALA513, GLY512, GLY519, ILE331, LEU103, LEU109, LEU338, LEU345, LEU370, LEU517, LEU511, MET99, MET508, PHE367, PHE504, SER339, SER516, TYR341, TYR371, TRP373, VAL102, VAL330, VAL335, VAL509 |
| | 1,7-bis(4-hydroxyphenyl)-1,6-heptadien-3,5-dione (active) | ARG106, ARG499, ALA513, GLY512, GLY519, HIS75, ILE503, LEU338, LEU345, LEU511, LEU517, MET99, PHE504, SER339, SER516, TYR334, TYR341, TRP373, VAL102, VAL330, VAL335, VAL509 |
| | 1,7-bis(4-hydroxy-3,5-dimethylphenyl)-1,6-heptadien-3,5-dione(active) | ARG106, ALA513, GLY512, GLY519, ILE98, LEU78, LEU338, LEU345, LEU370, LEU511, LEU517, MET508, PHE343, PHE367, PHE504, SER339, SER516, TYR101, TYR334, TYR341, TYR371, VAL74, VAL102, VAL330, VAL335, VAL509 |
| | 1,7-bis(3,5-diethyll-4-hydroxyphenyl)-1,6-heptadien-3,5-dione(active) | ARG106, ALA513, GLY512, GLY519, ILE98, LEU78, LEU338, LEU345, LEU370, LEU511, LEU517, MET508, PHE343, PHE367, PHE504, SER339, SER516, TYR101, TYR334, TYR341, TYR371, VAL74, VAL102, VAL330, VAL335, VAL509 |
| | 1,7-bis(2-methoxyphenyl)-1,6-heptadien-3,5-dione(active) | ARG106, ALA513, GLY512, GLY519, HIS75, ILE331, LEU103, LEU109, LEU338, LEU345, LEU370, LEU511, LEU517, MET99, MET508, PHE367, PHE504, SER339, SER516, TYR334, TYR341, TYR371, TRP373, VAL330, VAL335, VAL509 |
| | 1,7-bis(4-chlorophenyl)-1,6-heptadien-3,5-dione(active) | ARG106, ALA513, GLY512, GLY519, LEU78, LEU338, LEU345, LEU370, LEU511, LEU517, MET508, PHE367, PHE504, SER339, SER516, TYR101, TYR341, TYR371, TRP373, VAL74, VAL102, VAL330, VAL335, VAL509 |
| | 1,7-bis(4-methylphenyl)-1,6-heptadien-3,5-dione(active) | ARG106, ALA513, GLY512, GLY519, ILE98, LEU78, LEU338, LEU345, LEU370, LEU511, LEU517, MET508, PHE367, PHE504, SER339, SER516, TYR101, TYR341, TYR371, TRP373, VAL74, VAL102, VAL330, VAL335, VAL509 |
| | 1,7-bis(3,4-dibenzyloksphenyl)-1,6-heptadien-3,5-dione(inactive) | ARG106, ARG499, ALA513, GLU506, GLU510, GLY512, GLY519, HIS75, ILE98, ILE331, LEU78, LEU103, LEU109, LEU338, LEU345, LYS344, LYS346, LEU370, LEU511, LEU517, LEU520, MET99, MET508, PRO71, PHE504, SER339, SER516, TYR101,TYR334, TYR341, TYR371, TRP373, VAL74, VAL102, VAL330, VAL335, VAL509 |

IOP Conf. Series: Journal of Physics: Conf. Series **835** (2017) 012008 doi:10.1088/1742-6596/835/1/012008

Amino acid residues that interacted with both of active and inactive representative compounds means amino acid residues had no effect or even a negative effect on the activity of active representative compounds. Amino acid residues that interactedwith the active representative compounds and did not interact with inactive representative compounds meant that the amino acid residues influenced the activity of COX-2 enzyme inhibition of active representativecompounds. Amino acid residues exclusively bound to the active compounds but not present in inactive compounds were as follows :GLY340 (Glycine 340), ILE503 (Isoleucine 503), PHE343 (Phenylalanine 343), and PHE367 (Phenylalanine 367). Ligands had a binding site with specific amino acids in the receptor. Ligand-receptor interaction occured because there were hydrogen bonds, Van Deer Waals bonding, or electrostatic interactions. Using PyMOL, bond distance between the ligand and amino acids can be known. Distance bond with the amino acid receptor ligands will affect the bond strength (affinity) of receptor-ligands. The smaller the distance of the bond are the greater affinity of the ligand-receptor. 3D visualization of active compounds were done on 1,7-bis(2-methoxyphenyl)-1,6-heptadien-3,5-dion, using PyMOL application, the bond between the ligand and amino acids can be known. The bond between the ligand and amino acids can be seen in Figure 1.

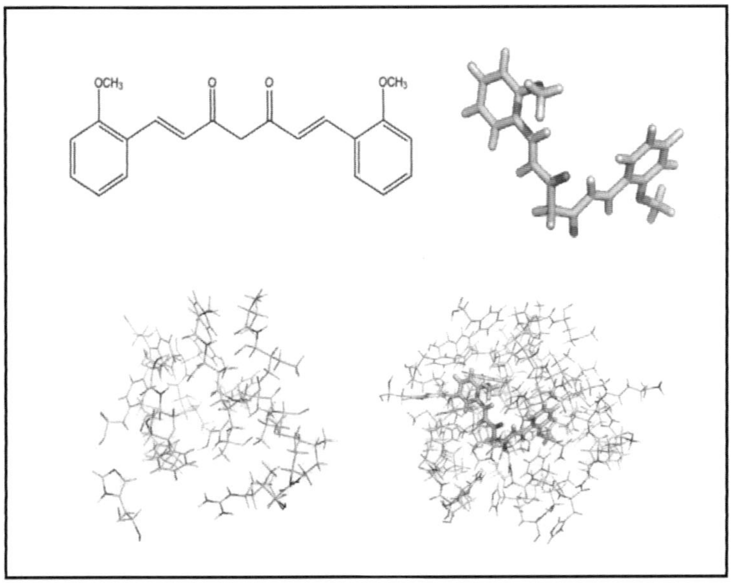


**Figure 1.** active representative compound 1,7-bis(2-methoxyphenyl)-1,6-heptadien-3,5-dion a) 2D model, b) 3D model c) bonding of compound on binding pocket COX-2

## 4. Conclusions

Of the 49 test compounds which were synthesized curcumin analogues compounds, there were eight compounds predicted active as COX-2 inhibitors using AYO_COX2_V .1.1 protocol i.e : 1,7-bis(3-methoxyphenyl)-1.6-heptadien-3,5-dione, 1,7-bis(4-hydroxyphenyl)-1.6-heptadien-3,5-dion, 1,7-bis (3-methoxyphenyl) -1.6-heptadien-3,5-dione, 1,7-bis(4-hydroxy-3,5-dimethylphenyl)-1.6-heptadien-3,5-dion, 1,7-bis(3,5-diethyl-4-hydroxyphenyl)-1.6-heptadien-3,5-dione, 1,7-bis (2-methoxyphenyl)-1.6-heptadien-3,5-dione, 1,7-bis(4-chlorophenyl)-1.6-heptadien-3,5-dione, and 1,7-bis (4-methylphenyl) -1,6-heptadien-3,5-dion and the active amino acids residues that increased the affinity of a COX-2 inhibitor i.e : GLY340 (Glycine 340), ILE503 (Isoleucine 503), PHE343 ( phenylalanine 343), and PHE367 (phenylalanine 367).

**References**

[1] Anand P, Thomas S G, Kunnumakkara A B, Sundaram C, Harikumar K B, Sung B, Tharakan S T, Misra K, Priyadarsini I K, Rajasekharan K N and Aggarwal B B 2008 *Biochem. Pharmacol.* **76**, 1590-611

[2] Mumpuni E, Widarsa A, Susilawati Y, Oisan, Nurrochmad A, Pranowo H D, Jenie U A dan Istyastono E P 2104 *JIFI* **12** 117-23

[3] Masuda T, Hidaka H, Shinohara A, Maekawa T, Takeda Y and Yamaguchi H 1999 *J Agric Food Chem.* **47** 71-77

[4] Nurfina A N, Reksohadiprodjo M S, Timmerman H, Jenie U A, Sugiyanta D and Vaner H G 1997 *Eur J Med Chem* **32** 321-328

[5] Sardjiman 2000 *Synthesis of some new series of curcumin analogues, antioxydative activity and structure activity relationship* [Disertasi] Yogyakarta Gadjah Mada University

[6] Supardjan A M and Da'i M 2005 *indonesian J. Pharm.* **16** 100-4

[7] Agarwal S, Reddy G V and Reddanna P 2009 *Expert Rev Clin Immunol* **5** 145-65

[8] Arief N 2004 *Rev Biofarmasi.* **2** 75-80

[9] Huang N, Shoichet B K and Irwin J J 2006 *J Med Chem* **49** 6789 – 801

IOP Conf. Series: Journal of Physics: Conf. Series **835** (2017) 012009 doi:10.1088/1742-6596/835/1/012009

# Genomic validation of PB 260 clone of rubber (*Hevea brasiliensis*) at Cikumpay Plantation by SSR marker

**J I Royani*, A Safarrida, I Rachmawati, H Khairiyah, I P Mustika, A Suyono, Y Rudiyana, Kubil, Nurjaya, and A Arianto**

Badan Pengkajian dan Penerapan Teknologi. Gd LAPTIAB 610, Kawasan PUSPIPTEK Serpong Tangerang Selatan Banten Indonesia 15314

E-mail: juwartina.ida@bppt.go.id

**Abstract.** Rubber from *Hevea brasiliensis* is the only commercial natural rubber in the world. Propagation of rubber trees usually done by grafting and seed germination. BPPT had been producing rubber tree by *in vitro* technique with embryo somatic methods. Validation of mother plant for *in vitro* propagation is important to compare between mother plant and propagated plants. The aim for this research was to validation of PB 260 clone that planted at Cikumpay Plantation by SSR marker. Sampling of 10 rubber leaves were done at Cikumpay Plantation based on GPS position from the area of PB 260 clone. Rubber leaves were isolated with CTAB modification method to obtained DNA. Four of SSR primers from rubber, i.e.: *hmac* 4, *hmac* 5, *hmct* 1, and *hmct* 5, were used as primers to amplification of rubber DNA. The result showed that no band that different from 10 rubber of PB 260 clone at Cikumpay Plantation. This research will continue to compare genomic validation between mother plant and propagated plants that had been produced from BPPT.

## 1. Introduction

Rubber from *Hevea brasiliensis* is the only one resource for commercial natural rubber in the world due to its high rubber content, good yield, and have the excellent physical properties of the rubber products and high quality [1]-[2]-[3]. Natural rubbers from *Hevea brasiliensis* cannot be replaced by synthetic alternatives because have unique properties, such as high elasticity, flexibility, resilience, impact and abrasion resistance, impermeable to liquids, efficient heat dispersion and malleability at cold temperature [4] which is difficult to be replaced by synthetic rubber in many applications

Now, Indonesia is the country with the largest rubber plantation in the world, almost 3.5 million hectares [5] but for productivity Indonesia still under Thailand with 26% or 3.2 thousand tons/year at 2016 [6]. Rubber in Indonesia grown by more than 1 million farmers, but most of plantation used of unselected seeds, almost 40% of all small holders' rubber plantations use of the low quality of seedlings. These made the production of rubber plantation in Indonesia still low.

Genomic validation with molecular marker now became the routine procedure to validate the plants and propagated plant especially when plants propagated by tissue culture and embryo somatic. Genomic validation for plants usually used for genetic diversity, authentication of plants, and genetic variation. For micropropagated plants, genomic validation used for identification and selection of mutation, confirm the uniformity, validation of somaclonal variation and epigenetic.

* Corresponding author

Molecular markers used for genomic validation were random amplified polymorphic DNA (RAPD), inter simple sequence repeat (ISSR), MSAP, amplified fragment length Polymorphism (AFLP), simple sequence repeat or microsatellite (SSR) and single nucleotide polymorphism (SNP) markers. Application of each methods based on availability, simplicity, cost, accurately, efficiency, highly reliable, and reproducibility [7]. The genomic validation of micropropagated plant by molecular marker had been reported by [8] for *Cassia angustifolia* Vahl with RAPD marker, [9] for *Dendrobium nobile* Lindl with RAPD and SCOT marker, [10] for *Nothapodytes nimmoniana* (Graham) Mabb with ISSR marker, [11] for *Brassica oleracea* L.var. italic with RAPD marker, [12] for *Pilosocereus robinii* with ISSR marker and [13]-[14] for *Hevea brasiliensis* with AFLP and SSR.

Validation of micropropagated plants, even by morphologic, genomic, pytochemical or with another character and method, will compared with mother plants to ensure that micropropagated plants true to type with mother plants. This point is very important to commercial utilization of miropropagated plants, especially for rubber clone.

The aim for this research was to validation of PB 260 clone of rubber tree at Cikumpay Plantation Purwakarta as a source of mother plants on micropropagated rubber at BPPT by genomic

## 2. Materials and methods

*2.1 Materials*

The materials used for this research were 10 of rubber leaves from PB 260 clone belongs to PT Perkebunan Nusantara VIII that planted at Cikumpay Plantations Purwakarta. Location of rubber tree sample by GPS position can see at Table 1.

**Table 1.** Location of rubber leaves sampling

| Sample | Location | | Altitude |
|---|---|---|---|
| | **(East)** | **(South)** | |
| 1 | 06°30.297' | 106°43.791' | 326 |
| 2 | 06°29.020' | 107°29.455' | 311 |
| 3 | 06°28.993' | 107°29.427' | 314 |
| 4 | 06°28.967' | 107°29.421' | 309 |
| 5 | 06°28.964' | 107°29.439' | 310 |
| 6 | 06°28.964' | 107°29.458' | 313 |
| 7 | 06°28.954' | 107°29.476' | 320 |
| 8 | 06°28.939' | 107°29.494' | 313 |
| 9 | 06°28.914' | 107°29.509' | 322 |
| 10 | 06°28.968' | 107°29.512' | 331 |

Primers used for this research were 4 primers based on GenBank with accession numbers are: AY135656, AY135657, AY135651 and AY135653 for the loci i.e. *hmac*4, *hmac*5, *hmct*1, *hmct*1, and *hmct*5 [15] as seen at Table 2.

*2.2 Sampling and DNA isolation*

Sampling of 10 rubber leaves from PB 260 clone was done at Cikumpay plantation Purwakarta based on GPS position. 10 rubber leaves were taken from 11 Ha of rubber plantation from PB 260 clone that planted since 2006. Young leaves were placed in to plastic clip, labeling and stored at cool box then bring to the Lab.

**Table 2.** SSR primers used for rubber DNA amplification [15]

| Primers | **Sequences** | **Size (bp)** |
|---|---|---|
| *hmac*4 | 5'-GTTTTCCTCCGCAGACTCAG-3' (L) | 216-272 |
| | 5'-ATCCACCAAATAAGGCATGA-3' (R) | |
| *hmac*5 | 5'-TCGGTTGGTTTACCATGACA-3' (L) | 270-286 |
| | 5'-ACATCACATGAGTGTATCTGATCTC-3'(R) | |
| *hmct*1 | 5'-AACCAGAAGGGTGTCATGCT-3' (L) | 196,198 |
| | 5'-GGAATCCCATGACAATCCAC-3' (R) | |
| *hmct*5 | 5'-ATGTATGTGTGCGCAGGAAG-3' (L) | 187-211 |
| | 5'-CTGTAGTCATGGCAGCAGGA-3' (R) | |

DNA isolation was done using a modified CTAB (*Cetyl trimethylammonium bromide*) method. The extraction buffer contains 10% CTAB, 5M NaCl, 0.5M EDTA pH 8.0, 1M Tris HCl pH 8.0 and purified water. Young leaves from PB 260 clone were weighted as much as 200 mg, cut out became slice and grounded with liquid nitrogen with mortar and pestle. Every sample was repeated 3 times for optimization method and collecting of DNA samples.

The purity and concentration of the extracted DNA were checked by electrophoresis with 1% gel agarose and measured by a Nanodrop spectrophotometer at 260/280 absorbance.

*2.3 Amplification of DNA rubber with SSR marker*

DNA amplification was running with composition as seen at Table 3. The condition for DNA amplification steps following a protocol described by [14] with initial denaturation step for 1 minute at 95°C, followed by elongation step with 35 cycles at 94°C for 30 s, 52°C for 60 s and 72°C for 120 s and a final elongation step at 72°C for 8 min. Composition of DNA amplification of rubber can see at Table 3. PCR products were visualization at electrophoresis with 1.2% gel agarose to analyze DNA bands.

**Table 3**. Composition of reagent for DNA amplification

| No. | **Reagent** | **Final Concentration** |
|---|---|---|
| 1. | Taq Polymerase Mix, 2X | 1 X |
| 2. | Forward Primer | 0,1-1,0μM |
| 3. | Reverse Primer | 0,1-1,0μM |
| 4. | Template DNA | 0,3 μM |
| 5. | Purified water | 0,3 μM |
| Total Volume | | 25μL |

*2.4 Analysis of DNA bands*

Analysis of DNA bands based on scoring from monomorphic and polymorphic bands that appears after visualization of PCR products on gel agarose. When DNA bands appear scored as absent (0) or present (1) to obtain binary data.

## 3. Results and discussions

*3.1 DNA isolation*

The result showed that isolation of DNA with modified CTAB successful to obtain DNA with the average of concentration between 1400-4000 μg/μL with quality of DNA was 1.15-2.05 at absorbance 260/280. Protocol from [16] said that DNA not contaminated with protein and RNA if the absorbance from 260/280 is 1.8-2.0. For this research, DNA with high purity was used as template for amplification.

### *3.2 DNA amplification with SSR marker*

DNA profile from 10 rubber trees from PB 260 clone had been amplified with 4 SSR primers from [15]. The result showed that all of samples have the same bands when amplified with each primer. Primer *hmac*4 has 1 single DNA band for each sample with size among 100-200 bp, this DNA bands different size with [15] that reported among 216-272 bp. For primer *hmac*5, the DNA profiles showed size among 200-300 bp with single band for each samples. They have same bands from 10 samples and same size with experiment reported by [15]. Primer *hmct*1 showed that DNA band from 10 sample of PB 260 clone have the same size and single band. This is different with [15] that reported with *hmct*1 primer, there were 2 alleles appear in the DNA profile. Amplification with primer *hmct*5 showed that DNA profiles from 10 samples of PB 260 clone have 2 alleles with size among 100-200 bp and 600-700 bp and have the same DNA bands.

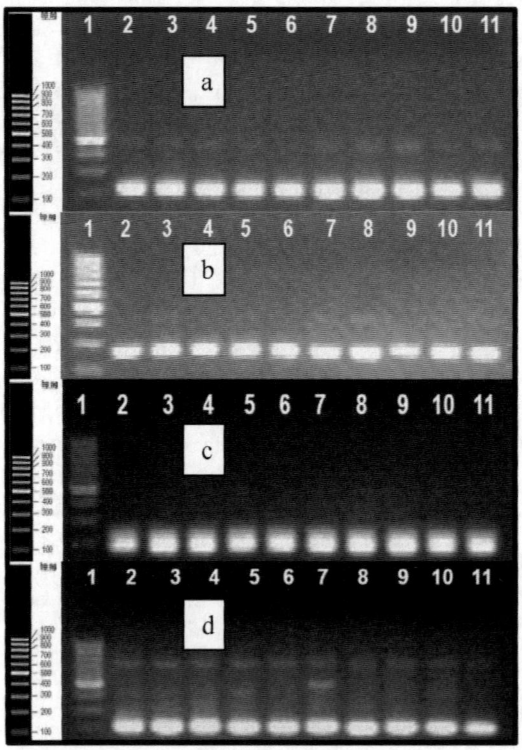


**Figure 1:** Genomic validation of 10 rubber trees from PB 260 clone, amplification with 4 SSR primers, (a) *hmac*4, (b) *hmac*5 (c) *hmct*1 and (d) *hmct*5

DNA profiles of 10 samples from PB 260 clone at Cikumpay Plantation showed that no different clone and no genetic change even have been planted since 10 years ago. This research ensures that PB 260 clone used as a source of mother plant on micropropagated plants at BPPT same as genetic. This is the first report of genomic validation of rubber tree after a long planted at rubber plantations. After certain that no change genetically of mother plant, this research will continue to compare genomic validation between mother plant and micropropagated plants that had been produced from BPPT.

## 4. Conclusions

4 SSR primers successful to amplified 10 rubber of PB 260 clone. Genetic validation of 10 rubber of PB 260 clone showed no different and genetic change of clone.

## References

[1] Hayashi Y 2009 *Plant Biotechnol. J.* **26** 67-70
[2] Venkatachalam P, PK Jayasree, S Sushmakumari, R Jayashree, K Rekha, S Sobha, P Priya, RG Kala, and Thulaseedharan A 2007 *Funct Plant Sci. Biotechnol.* **1** 1-17
[3] Asawatreratanakul K, YW Zhang, D Wititsuwannakul, R Wititsuwannakul, S Takahashi, A Rattanapittayaporn and Koyama T 2003 *Eur. J. Biochem.* 270 4671-80
[4] Cornish K 2001 *Phytochem.* **57** 1123-34
[5] Dirjenbun. *Kebijakan tanaman karet di Indonesia*. Dalam: Focus Group Discussion Pengembangan Teknologi Perbanyakan Bibit Unggul dan Pengendalian Penyakit dalam Rangka Meningkatkan Kualitas dan Produktivitas Tanaman Karet, BPPT. Serpong, 02 Oktober 2013.
[6] Karet (2016, October 16) Retrieved from http://www.indonesia-investments.com/id/bisnis/komoditas/karet/item185.
[7] Semagn K, A Bjørnstad, and Ndjiondjop MN 2006 *Afr. J. Biotechnol.* **5** 2540-68
[8] Chetri SK, PR Sardar and Agrawal V 2014 *Physiol Mol Biol Plants* **20** 517-26
[9] Bhattacharyya P, S Kumaria, R. Diengdoh, and Tandon P 2014 *Meta Gene* **2** 489-504
[10] Prakash L, SK Middha, SK Mohanty, and Swamy MK 2016 *3 Biotech* **6** 171
[11] Kumar P, G Gambhir, A Gaur and Srivastava DK 2015 *Curr. Sci.* **109** 1470-5
[12] Khattab S, FE Sherif, HA El-Garhy, S Ahmed, and Ibrahim A 2014 *Gene* **533** 313-21
[13] Medina C, I García, M Caro, and Aristizábal FA 2007 *Rev. Col. Cienc. Quím. Farm* **36** 70-80
[14] Srichuay, W and Te-chato S 2015 *Songklanakarin J. Plant Sci.* **2** 27-31
[15] Saha T, C Bindu Roy, and Nazeer MA 2005 *Plant Breeding* **124** 86 –92
[16] Sambrook J, EF Fritsch, and Maniatis T 1989 *Molecular cloning: a laboratory Manual* 3 (New York, Cold Spring Harbor Laboratory Press, Cold Spring Harbor)

## Acknowledgment

We would like to say thank to Management of Cikumpay Plantation PT Perkebunan Nusantara VIII Purwakarta who permit us for sampling and to Deputy of Technology for Agroindustry and Biotechnology BPPT for supporting this research.

IOP Conf. Series: Journal of Physics: Conf. Series **835** (2017) 012010 doi:10.1088/1742-6596/835/1/012010

# Indonesian palm civet coffee discrimination using UV-visible spectroscopy and several chemometrics methods

**M Yulia[1,*] and D Suhandy[2]**
[1] Department of Agricultural Technology, Lampung State Polytechnic, Jl. Soekarno Hatta No. 10, Rajabasa Bandar Lampung, Indonesia
[2] Department of Agricultural Engineering, Faculty of Agriculture, The University of Lampung, Jl. Prof. Dr. Soemantri Brojonegoro No.1, Bandar Lampung, 35145, Indonesia

E-mail: meinilwitayulia@polinela.ac.id

**Abstract**. Indonesian palm civet coffee or kopi luwak (Indonesian words for coffee and palm civet) is well known as the world's priciest and rarest coffee. To protect the authenticity of luwak coffee and protect consumer from luwak coffee adulteration, it is very important to develop a simple and inexpensive method to discriminate between civet and non-civet coffee. The discrimination between civet and non-civet coffee in ground roasted (powder) samples is very challenging since it is very difficult to distinguish between the two by using conventional method. In this research, the use of UV-Visible spectra combined with two chemometric methods, SIMCA and PLS-DA, was evaluated to discriminate civet and non-civet ground coffee samples. The spectral data of civet and non-civet coffee were acquired using UV-Vis spectrometer (Genesys™ 10S UV-Vis, Thermo Scientific, USA). The result shows that using both supervised discrimination methods: SIMCA and PLS-DA, all samples were correctly classified into their corresponding classes with 100% rate for accuracy, sensitivity and specificity, respectively.

## 1. Introduction

The use of infrared spectroscopy both NIR (near infrared) and MIR (mid infrared) as a rapid and non-destructive technique for the authentication and adulteration detection of food samples has been reported by several researchers. NIR spectroscopy utilizes the spectral range from 700 to 2500 nm and provides much more complex structural information related to the vibrational transitions associated mainly with C-H, N-H, and O-H functional groups. NIR spectroscopy has been used to detect food authentication in honey [1], milk powder [2], fishmeal [3] and coffee [4]. Recent researches reported effective detection of food adulteration using NIR spectroscopy in honey [5], coffee [6] and meat [7].

In MIR region (2500-25000 nm), spectra are commonly used for structural identification (fingerprinting) of organic compounds because the absorption bands are caused by fundamental vibrations of a specific functional group [8]. Authentication of food in this region has been reported for extra virgin olive oil [9], artisanal honeys adulterated with sugar solutions [10] and lard adulteration in cake [11].

In Raman spectroscopy, several researches for food authentication have been published. For example, rapid quantitative assessment of the adulteration of virgin olive oils with hazelnut oils using

[*] Corresponding author

Raman spectroscopy and chemometrics was reported with the levels of hazelnut oils used to adulterate extra virgin olive oil were successfully quantified using partial least squares and genetic programming [12].

While there has been a steady growth in food authentication applications for NIR and MIR spectroscopy, there has been very little research conducted using UV–Visible spectroscopy (200-700 nm) for food authentication. Recently, the use of UV-Visible spectroscopy for identification of adulteration in ground roasted coffee has been reported with a 100% classification was obtained with SPA-LDA variable selection [13]. Using spectral region in the range 200-400 nm combined with SIMCA, a feasibility study for separation between civet and non-civet coffee was demonstrated [14]. Comparing to NIR and MIR spectroscopy, spectroscopy in UV-Visible region has an advantage in fact that spectrometer in this region is relatively low cost and it is available to most standard laboratories. This method is also a green technology without chemical waste during sample extraction.

Civet coffee or kopi luwak is one of the most rarest and expensive coffee in the world. It is one of the Indonesian coffee specialties. Civet coffee is any coffee bean (arabica or robusta) which has been eaten and passed through the digestive tract of Asian palm civet (*Paradoxurus hermaphroditus*), which uses its keen senses to select only the best and ripest berries. Due to the high price and limited production, civet coffee has been one the target for food adulteration. Several types of cheap non-civet coffees have been added into pure civet coffee to lead mislabeling of authentic civet coffee. Up to now, there is no research on spectroscopy area for civet coffee authentication. Therefore, in this study we reported an evaluation of using UV-Visible spectroscopy in combination with SIMCA (soft independent modelling of class analogies) and PLS-DA (partial least square-discrimination analysis) for discrimination between civet and non-civet coffee.

## 2. Research methodology

### *2.1. Samples*

A number of 1 kg roasted robusta civet coffee (Indonesian palm civet coffee) and 1 kg roasted non-civet robusta coffee (Pagar Alam robusta coffee) were collected directly from coffee farmers at Liwa, Lampung, Indonesia (Hasti coffee Lampung). All coffee samples were grinded using home-coffee-grinder (Sayota). To avoid significant influence of particle sizes in coffee powder to spectral analysis, it is important to use same particle size in coffee powder samples [15]. In this research we use particle size of 420 μm by sieving through a nest of U. S. standard sieves (Mesh number of 40) on a Meinzer II sieve shaker (CSC Scientific Company, Inc. USA) for 10 minutes. The experiments were performed at room temperature (around 27-29°C). We prepare 100 samples of coffee samples which consist of two classes; class A and B. Class A has 50 samples of civet coffee without adulteration (pure 100% only civet coffee). Class B has 50 samples of robusta Pagar Alam coffee (non-civet coffee). The samples were divided into two groups, calibration set (60 samples) and prediction set (40 samples), randomly. Calibration both in SIMCA and PLS-DA was done with full-cross validation method. Number of optimal principal components (PCs) in SIMCA model and latent variables (LVs) in PLS-DA model was determined with the lowest root mean square error of cross-validation (RMSECV).

### *2.2. Extraction procedures*

In this research, samples in the solution form were used for spectral acquisition. For this, an aqueous extraction procedure of the coffee samples was performed as described by Souto et al. [13]. First, 1.0 g of each sample was weighed and placed in a glass beaker. Then, adding 10 mL of distilled water at 90-98°C then mixed with magnetic stirring (Cimarec™ Stirrers, model S130810-33, Barnstead International, USA) at 350 rpm for 5 min. Then the samples were filtered using a 25 mm pore-sized quantitative filter paper coupled with an erlenmeyer. After cooling process to room temperature (for 20 min), all extracts were then diluted in the proportion of 1:20 (mL: mL) with distilled water. UV-Visible spectra from the aqueous extracts were acquired using a UV-Vis spectrometer (Genesys™ 10S UV-Vis, Thermo Scientific, USA).

IOP Conf. Series: Journal of Physics: Conf. Series **835** (2017) 012010 doi:10.1088/1742-6596/835/1/012010

### *2.3. Spectral measurement*

UV-Vis spectra in the range of 200-400 nm were acquired by using a UV-Vis spectrometer (Genesys™ 10S UV-Vis, Thermo Scientific, USA) equipped with a quartz cell with optical path of 10 mm, and spectral resolution of 1 nm at a room temperature. This is a dual beam spectrometer with sample and reference holder are separated each other. Before the measurements step, reference spectra was measured by putting 3 mL of distilled water. It was followed by sample measurement by putting 3 mL of coffee aqueous samples. The relative absorbance of each sample was calculated and then stored using USB flash disk.

### *2.4. Multivariate analysis using SIMCA and PLS-DA*

All recorded spectra data were transferred to computer via USB flash disk and then convert the spectra data from .csv extension into an excel data (.xls). The original (raw) spectra were used for further analysis. For classification and discrimination, two different multivariate analyses were used, a SIMCA (based on PCA algorithm) and PLS-DA (based on PLS algorithm). The calculation of SIMCA and PLS-DA model was done using a multivariate analysis program The Unscrambler® version 9.8 (CAMO AS, Norway). Sensitivity (samples belonging to the class and classified correctly in this class) and specificity (samples not belonging to the modelled class and correctly classified as not belonging) were considered for the evaluation of the classification achieved with the multivariate methods [16]. The sensitivity and specificity were calculated according to equations:

$$\text{Sensitivity} = \frac{\text{TP}}{\text{TP} + \text{FN}} \tag{1}$$

$$\text{Specificity} = \frac{\text{TN}}{\text{TN} + \text{FP}} \tag{2}$$

where TP is the number of true positives, FN is false negatives, TN is true negatives and FP is false positives.

## 3. Results and discussions

### *3.1. Typical spectra of civet and non-civet coffee samples*

Figure 1 demonstrated a typical spectra of civet and non-civet coffee in the range 200-400 nm. It can be seen that spectra of civet and non-civet coffee are almost similar with the positive and negative peaks located at 260 nm, 280 nm, 310 nm and 325 nm. The sharp negative peak at 260 nm and the positive peak at 280 nm may be related with the absorbance of caffeine which is also reported by several previous reports [17], [18].

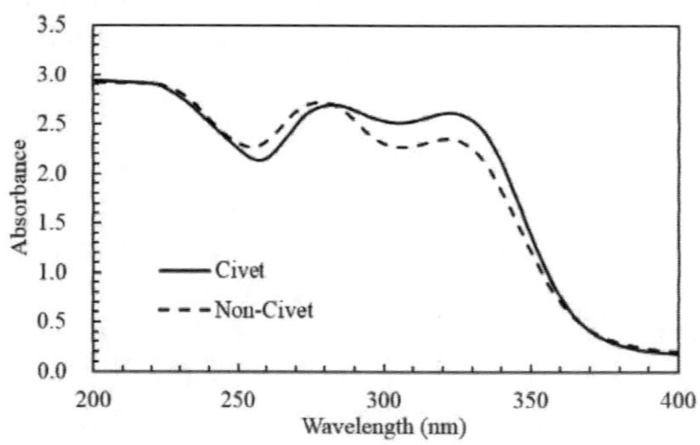


**Figure 1.** Typical spectra of civet and non-civet coffee samples.

### *3.2. Unsupervised discrimination using PCA*

The unsupervised algorithm applied was principal component analysis (PCA). PCA analysis shows score plot of PC1 and PC2. Here we can see that the coffee samples can be clustered into two groups, civet (red marker) and non-civet coffee (blue marker) (as shown in Figure 2). The use of PC1 and PC2 can explain 98% of total variance in the coffee samples. Civet coffee is clustered in the right of PC1 while non-civet coffee is clustered on the left one. PCA results suggest that it is possible to discriminate between civet and non-civet coffee.

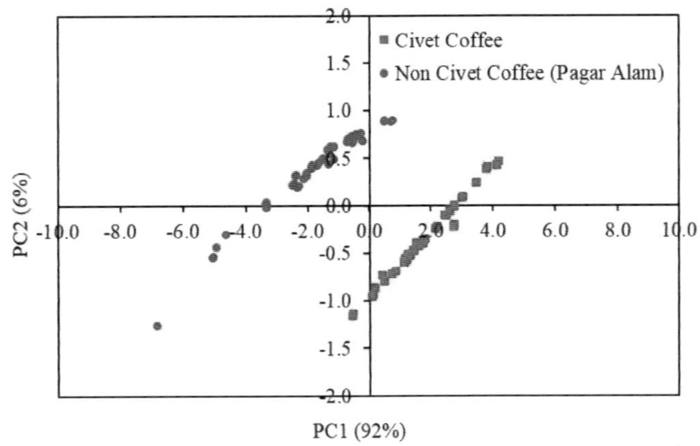


**Figure 2.** Score plot of PC1 and PC2 of coffee samples.

### *3.3. Supervised methods using SIMCA and PLS-DA*

For SIMCA, each SIMCA model was developed for each corresponding class, civet and non-civet coffee using calibration samples in each class (30 samples for civet and 30 samples for non-civet, respectively). Using the developed SIMCA model, discrimination of prediction samples was applied and the result was presented in Table 1. All analysed samples were classified correctly in relation to sensitivity and specificity, with results of 100%.

**Table 1.** Discrimination results between civet and non-civet samples in prediction samples using the soft independent modelling of class analogy (SIMCA) method.

| Spectral data | Samples | Classified correctly to proper class | Classified to none | Classified to both | Number of PCs in model |
|---|---|---|---|---|---|
| Original | Civet | 20 | 0 | 0 | 3 |
| | Non Civet | 20 | 0 | 0 | 1 |
| | Total | 40 | 0 | 0 | |

Using PLS-DA, a calibration model based on PLS was developed using calibration sample set (in total 60 samples, 30 civet and 30 non-civet samples). In this PLS model, wavelengths (200-400 nm) are the predictors variable and type of coffee (civet or non-civet) is target or response variable. Since that type of coffee must be quantified in a numeric value, a dummy value for type of coffee was assigned with 1 = civet coffee and 2 = non-civet coffee. A threshold value of 0.5 is adopted, so that, if y is above this limit, the sample is assigned to class 2 (here, non-civet coffee), while if it is below, it is assigned to class 1 ("civet coffee") [19] [20]. The PLS-DA model was obtained with 2 latent variables (LVs). It is also important to note that the agreement between the root mean square error of calibration (RMSEC) and RMSECV, 0.043 and 0.045, respectively, confirms the absence of overfitting in the developed PLS-DA calibration model. Using this PLS-DA model, a prediction of type of coffee was applied using prediction sample set (40 samples). The result was presented in Table 2. Similar to SIMCA prediction result, all samples are belongs to each proper corresponding class and resulted in 100% rate for sensitivity and specificity. This result shows that SIMCA and PLS-DA method could be effective to discriminate between civet and non-civet coffee.

**Table 2.** Discrimination results between civet and non-civet samples using the partial least square-discriminant analysis (PLS-DA) method.

| Spectral data | Samples | Classified correctly to proper class | Classified to none | Classified to both | Number of LVs in model |
|---|---|---|---|---|---|
| Original | Civet | 20 | 0 | 0 | 2 |
| | Non Civet | 20 | 0 | 0 | 2 |
| | Total | 40 | 0 | 0 | |

## 4. Conclusion
In this research we demonstrate a feasibility of using UV-Visible spectroscopy along with chemometrics method to discriminate civet and non-civet coffee. The classification result using SIMCA and PLS-DA methods was satisfactory with sensitivity and specificity was 100%, respectively. This result can be used for developing a possible application of using UV-Visible spectroscopy and chemometrics method to establish a rapid and acceptable method for authentication of Indonesian civet coffee in the near future.

## References
[1] Ruoff K, Luginbuhl W, Bogdanov S, Bosset J O, Estermann B, Ziolko T and Amado R 2006 *J. Agric. Food Chem.* **54** 6867–72.
[2] Maraboli A, Cattaneo T M P and Giangiacomo R 2002 *J. Near Infrared Spec.* **10** 63–9.
[3] Murray I, Aucott A S and Pike I 2001 *J. Near Infrared Spec.* **9** 297–311.
[4] Downey G, Boussion J and Beauchene D 1994 *J. Near Infrared Spec.* **2** 85–92.
[5] Bázár G, Romvári R, Szabó A, Somogyi T, Éles E and Tsenkova R 2016 *Food Chem.* **194** 873–80.
[6] Bona E, Marquetti I, Link J V, Makimori G Y F, da Costa Arca V, Guimarães Lemes A L, Ferreira J M G, dos Santos Scholz M B, Valderrama P and Poppi R J 2017 *LWT - Food Sci. Technol.* **76** 330–6.
[7] Alamprese C, Casale M, Sinelli N, Lanteri S and Casiraghi E 2013 *LWT - Food Sci. Technol.* **53** 225–32.
[8] Guillen M and Cabo N 1997 *J. Sci. Food Agr.* **75** 1–11.
[9] Gurdeniz G and Ozen B 2009 *Food Chem.* **116** 19–25.
[10] Kelly J, Petisco C and Downey G 2006 *J. Agric. Food Chem.* **54** 66–71.
[11] Syahariza Z A, Che Man Y B, Selamat J and Bakar J 2005 *Food Chem.* **92** 65–71.
[12] López-Díez E C, Bianchi G and Goodacre R 2003 *J. Agric. Food Chem.* **51** 6145–50.
[13] Souto U T C P, Barbosa M F, Dantas H V, Pontes A S, Lyra W S, Diniz P H G D, Araújo M C U and Silva E C 2015 *LWT - Food Sci Technol.* **63** 1037–41.
[14] Suhandy D, Yulia M, Waluyo S, Sugianti C, Iriani R, Handayani F N and Apratiwi N 2016 *Proc. of the USR International Seminar on Food Security* (Bandar Lampung: Universitas Lampung) pp 245–53.
[15] Suhandy D, Waluyo S, Sugianti C, Yulia M, Iriani R, Handayani F N and Apratiwi N 2016 *Proc. of Seminar Nasional Tempe* (Bandar Lampung: Universitas Lampung) pp 198–204.
[16] Grasel F S and Ferrao M F 2016 *Anal. Methods.* **8** 644–9.
[17] Belay A, Ture K, Redi M and Asfaw A 2008 *Food Chem.* **108** 310–5.
[18] Clarke R J and Macrae R 1985 *Coffee chemistry (Vol. 1)* (Amsterdam: Elsevier)

[19] Ståhle L and Wold S 1987 *J. Chemom.* **1** 185–96.
[20] Barker M and Rayens W 2003 *J. Chemom.* **17** 166–73.

**Acknowledgment**

We thank support of the Indonesian Ministry of Research, Technology and Higher Education (KEMENRISTEKDIKTI) for financially support this research through Penelitian Strategis Nasional (STRANAS) 2016 (Nomor: 419/UN26/8/LPPM/2016) and The University of Lampung, Indonesia. We also thank Hasti Coffee Lampung to prepare the samples and thank to Department of Agricultural Engineering, The University of Lampung for permission to use spectrometer and its supporting devices.

# Combination of near infrared spectroscopy and chemometrics for authentication of taro flour from wheat and sago flour

**Rachmawati[1], E Rohaeti[1*] and M Rafi[1,2]**

[1]Department of Chemistry, Faculty of Mathematics and Natural Sciences, Bogor Agricultural University, Kampus IPB Darmaga Indonesia
[2]Halal Science Centre, Bogor Agricultural University, Kampus IPB Baranangsiang Jalan Raya Pajajaran Bogor 16151, Indonesia

E-mail : etirohaeti@ ipb.ac.id

**Abstract.** Taro flour on the market is usually sold at higher price than wheat and sago flour. This situation could be a cause for adulteration of taro flour from wheat and sago flour. For this reason, we will need an identification and authentication. Combination of near infrared (NIR) spectrum with multivariate analysis was used in this study to identify and authenticate taro flour from wheat and sago flour. The authentication model of taro flour was developed by using a mixture of 5%, 25%, and 50% of adulterated taro flour from wheat and sago flour. Before subjected to multivariate analysis, an initial preprocessing signal was used namely normalization and standard normal variate to the NIR spectrum. We used principal component analysis followed by discriminant analysis to make an identification and authentication model of taro flour. From the result obtained, about 90.48% of the taro flour mixed with wheat flour and 85% of taro flour mixed with sago flour were successfully classified into their groups. So the combination of NIR spectrum with chemometrics could be used for identification and authentication of taro flour from wheat and sago flour.

## 1. Introduction

Taro (*Colocasia esculenta* (L.) Schott) or locally known as *talas* in Indonesia is famous and widely used for food. Many types of food such as cake in traditional or modern form were derived from taro flour. In the market, taro flour is usually sold at higher prices compare to the other kind of flour such as wheat and sago flour. The price of taro flour is about Rp25 000/kg while for wheat and sago flour is about Rp10,000-15,000/kg. With this situation, taro flour could be adulterated with wheat and sago flour with a purpose to reduce the production costs and also because of the availability of wheat and sago flour more available than taro flour. The substitution and mixing with other similar material could decrease the quality of the end product of taro flour. To prevent this thing we must have an authentication method for quality control of taro flour to guarantee the authenticity of taro flour.

Authentication of food raw materials is a part in the quality control of final food product will require an adequate analytical method. In the previous studies, some researcher has been demonstrated the used of near infrared spectroscopy (NIR) combined with chemometrics in the development of authentication methods for quality control purposes. NIR technique could be utilized in the

[*] Corresponding author

IOP Conf. Series: Journal of Physics: Conf. Series **835** (2017) 012011 doi:10.1088/1742-6596/835/1/012011

development of authentication method because able to detect small differences in the chemical composition present in some material. Also, this technique is relatively fast, requires no sample preparation, didn't use chemicals, non-destructive measurement and relatively cheaper [1,2]. NIR spectrum is a complex pattern, so direct and visual interpretation become not easy if we would like to use to differentiate a similar or mixed material to its constituent. Chemometrics could be used to facilitate the interpretation of data by using multivariate analysis i.e. principal component or discriminant analysis for the purposes [3,4]. This combination technique has been used for rapid estimation of the quality of taro [5], evaluation ability of NIR spectroscopy to authenticate green asparagus as a function of the growth [6] and roasting degree of coffee [7]. In this study, we have been developed an authentication method of taro flour from wheat and sago flour by NIR and chemometrics analysis.

## 2. Materials and Methods

### *2.1. Material*
Samples of taro were collected from several regions, two from Bogor (Cibereum and Ciapus), one from Cianjur, Bandung, Kuningan, Cimahi, Sumedang and one commercial taro flour. Five wheat and four sago flour were purchased from local market in Bogor.

### *2.2. Sample preparation*
Taro from several regions is prepared to powdered form according to the method described by Miranda et al. [8]. Firstly, we cleaned the taro from the soil and then washed with water. After the taro was drained, we peeled and washed again, and then sliced about 0.2 cm thickness. The sliced taro dried at $65^0$C for 5 hours in the oven. Dried sample then pulverized with a grinding machine and sieved to obtain a 100 mesh particle size.

### *2.3. NIR spectrum measurement*
Into the petri dish, we added the sample until reach ¾ height of the cup. All of the samples (taro flour 100%, taro flour 95% + wheat flour 5%, taro flour 75% + wheat flour 25%, taro flour 50% + wheat flour 50% and wheat flour 100%) were weighed and the same composition also used for sago flour. NIR spectrum was acquired with NIRFlex Solids Petri N-500 spectrophotometer (Buchi, Flawil, Switzerland) with attenuated total reflection mode the near infrared region (10000–4000 cm$^{-1}$) with 32 scans/min and resolution of 4 cm$^{-1}$.

### *2.4. Chemometrics analysis*
Preprocessing of spectral data is an important step before subjecting the NIR spectra data for multivariate analysis. We used normalization and standard normal variate (SNV) for signal processing of the NIR spectra. Principal component analysis (PCA) and discriminant analysis (DA) were used to build an authentication model using XLSTAT software version 2012 (Addinsoft, New York, USA).

## 3. Result and Discussions

### *3.1. NIR Spectrum*
NIR spectrum of all taro, wheat and sago flour samples used in this work give same profile and only differ in the absorbance value under the same measurement conditions (Figure 1). This result is indicating that the chemical compounds present in the samples are same and only differ in the concentration. From the NIR spectrum of taro flour, we found eight characteristic peaks. Peak appear in the 1450 and 1540 nm correspond to stretching vibration of O—H in the first overtone. The peak at 1930 nm denotes stretching vibration of O—H or a combination of deformation H-O-H, while the peak at 1960 nm assigned as a vibration of O—H or a combination of bending vibration. A peak at 2100 nm indicates bending vibration of O—H or a combination of stretching vibration of C—O. A peak at 2280

and 2322-2330 nm defined as stretching vibration of C—H or deformation of CH2, while for peak appear in 2500 nm denotes stretching vibration of C—H or C—C or C—O [1].

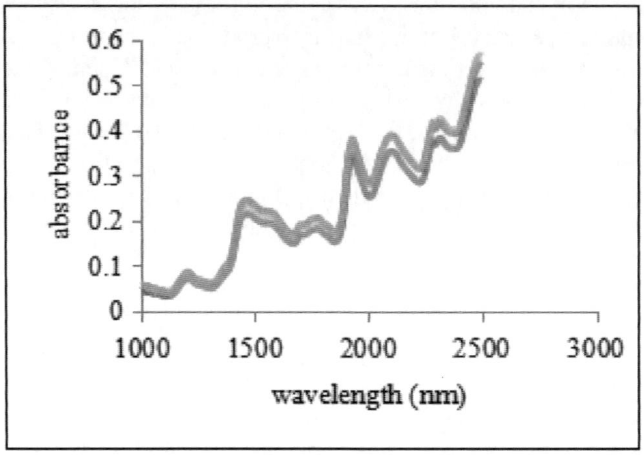


**Figure 1.** Representation of NIR spectra of taro flour from Kuningan (—), Bogor Ciapus (—) and Sumedang (—)

Based on the results of taro flour NIR spectra, we selected three samples from Bogor-Ciapus, Sumedang, and Kuningan that show high, medium and low absorption to build an authentication method of taro flour. The three samples were mixed with wheat and sago flour as counterfeit material to taro flour are. Figure 2 showed the NIR spectra of the mixture samples (5% wheat and sago flour added to taro flour) give same pattern and only differ in the intensity.

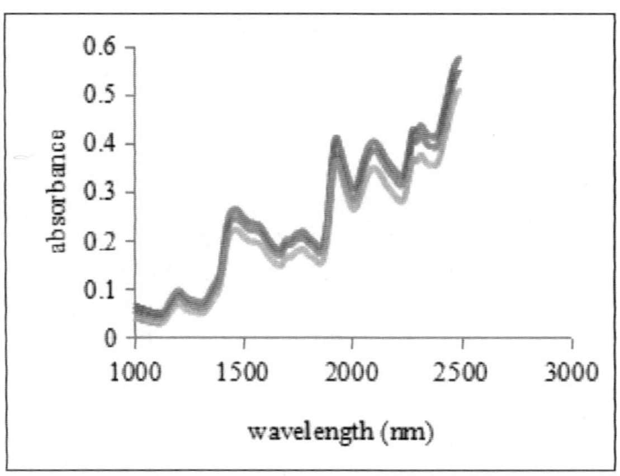


**Figure 2.** Representation of NIR spectra of taro flour (—), wheat flour (5%) in the taro flour (—), sago flour (5%) in the taro flour (—)

*3.2. Identification and authentication taro flour from wheat and sago flour*
Distinguishing the three flour samples is crucial to identify and authenticate the taro flour. The combination of NIR spectra and chemometrics could be used as the potential choice for such purposes. Some preprocessing signal on the NIR spectra of the sample should be conducted to obtain a good

predictive authentication model of the samples used. Preprocessing signal can avoid some problems due to baseline drift and reduce random noise in the initial spectrum. This process would lead to the distinctive character of the spectrum becomes more quantized, so the identifier factors become increasingly precise.

3.2.1. *Principal Component Analysis.* PCA is one multivariate analysis widely used for a purpose to reduce data and extract information to find a combination of variables describing major trends in a data set. Each new variable knows as a principal component (PC) that is generated as a linear combination of the original variables measurement. In PCA we will have first principal components (PC1) which have the greatest variance and second principal component (PC2) which has the next largest variance in the data set [4]. Figure 3 showed the PCA plot by using the first two PCs with explaining 96.62% of the total variance (PC-1 = 87.83%, PC-2 = 8.79%). Based on the PCA plot, we found that the three flour samples were separated according to the type of flour. So, PCA was able to distinguish the taro, wheat and sago flour.

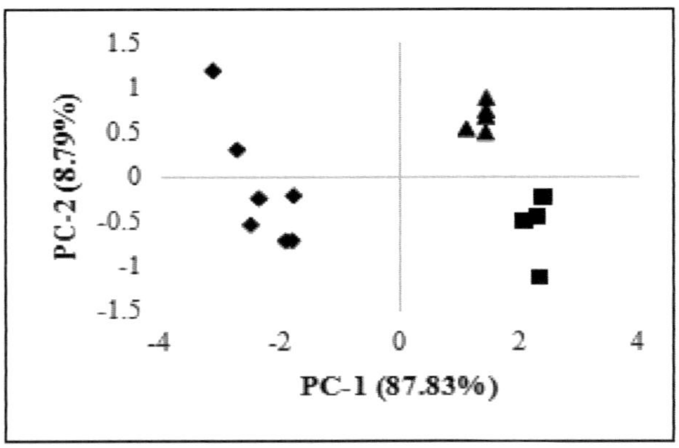


**Figure 3.** PCA plot of samples: taro flour (♦), wheat flour (▲), and sago flour (■)

We also used PCA to build an authentication model of taro flour when mixed with wheat or sago flour. PCA generates a first two PCs explain 95.86% of the total variance (PC-1 = 86.28%, PC-2 = 9.58%) for the sample mixed with wheat flour (5, 25 and 50%) and 95.58% of the total variance (PC-1 = 87.57%, PC-2 = 8.01%) for the sample is mixed with sago flour (5, 25 and 50%) (Figure 4 and 5). Based on the result obtained, separation of the pure taro flour and mixed with wheat or sago flour was achieved but didn't give clear separation. To have better separation, we used another multivariate analysis namely discriminant analysis.

3.2.2. *Discriminant Analysis.* DA is one of the supervised pattern recognition techniques and work to find a discriminant factor (DF) for each group by finding a linear combination of data provide separation of two or more groups of observations [9]. In this work, we used the first two PC obtained from PCA to build a predictive model for authentication of taro flour. The use of the first two PC because the PCA plot of this two PCs could distinguish well the three samples. From the DA we obtained two DF with total variance greater than 99% (DF-1 = 98.79% and DF-2 = 1.21%). Separation of the three samples by DA is much better because the distance is much longer than the PCA. Evaluation of the predictive ability of the model to distinguish the three sample by cross-validation also gives a satisfactory result, about 100% of the samples are identified according to its type.

IOP Conf. Series: Journal of Physics: Conf. Series **835** (2017) 012011 doi:10.1088/1742-6596/835/1/012011

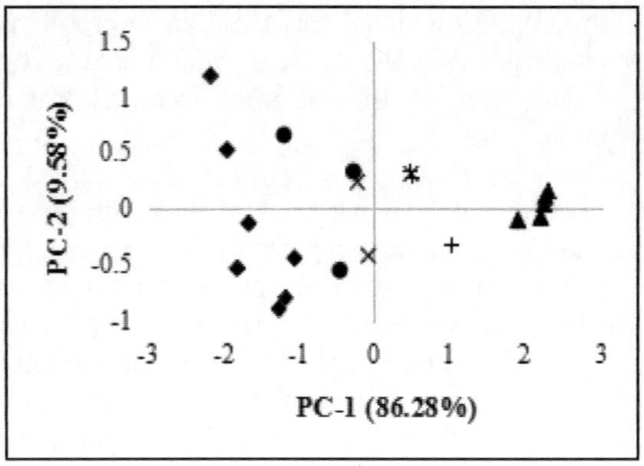


**Figure 4.** PCA plot of samples: taro flour (♦), wheat flour (▲), 5% (●), 25% (×), 50% (+) wheat flour in taro flour

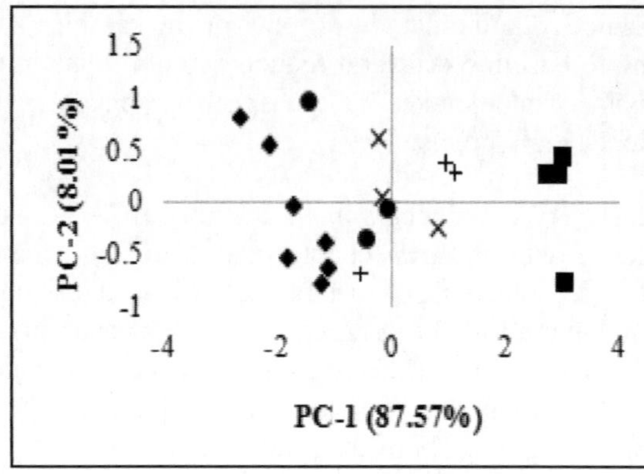


**Figure 5.** PCA plot of samples: taro flour (♦), sago flour (■), 5% (●), 25% (×), 50% (+) sago flour in taro flour

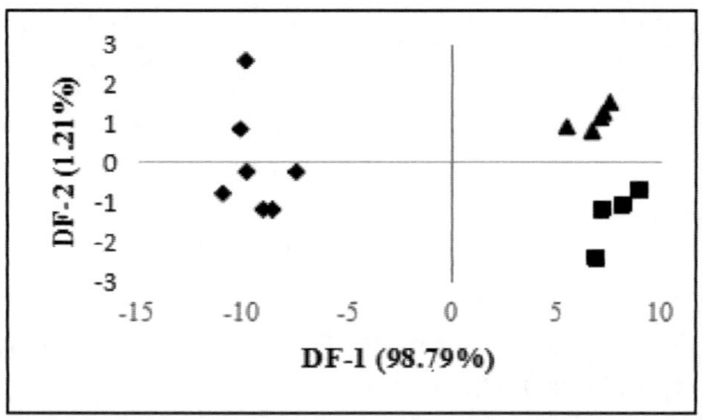


**Figure 6.** DA plot of samples: taro flour (♦), wheat flour (▲), sago flour (■)

The first two PC is also used to create an authentication model using DA for mixed taro flour with wheat and sago flour as the counterfeiters. In the plot of DA from taro flour mixed with wheat flour and sago flour we found the total variance of the two DF were 100%. From the DA plot (Figure 7 and 8) the separation of pure taro flour with the mixed taro with wheat and sago flour (5, 25 and 50%) is much clearer compare with PCA. The evaluation of the predictive ability of the model by cross-validation give an adequate accuracy with 90.48% and 85% of the samples are mixed with wheat and sago flour respectively were successfully classified into their group. Based on the results of DA, we could authenticate the taro flour by using the NIR spectra follow with DA.

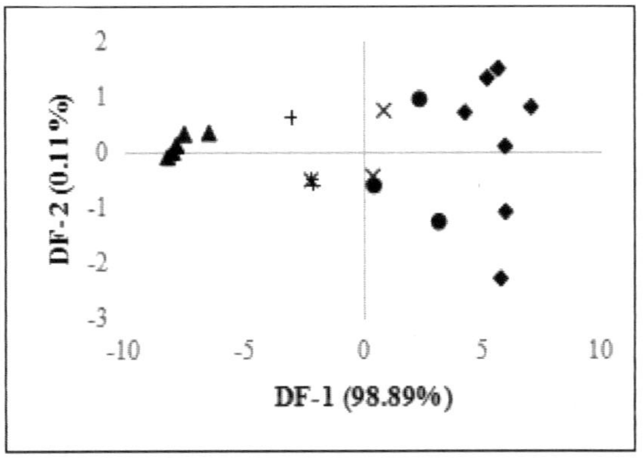


**Figure 7.** DA plot of samples: taro flour (♦), wheat flour (▲), 5% (●), 25% (×), 50% (+) wheat flour in taro flour

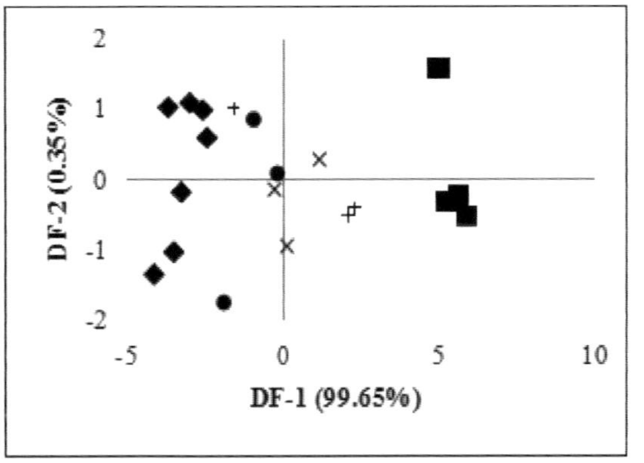


**Figure 8.** DA plot of samples: taro flour (♦), sago flour (■), 5% (●), 25% (×), 50% (+) sago flour in taro flour

## 4. Conclusions

The combination of NIR spectra and chemometrics could be used for authentication of taro flour from wheat and sago flour. PCA and DA could be utilized for the authentication of taro flour from wheat and sago flour. DA give clearer separation of pure taro flour (100%) and taro flour mixed with 5, 25 and 50% of wheat and sago flour.

## References

[1] Burns D and Ciurczak E 2008 *Handbook of Near-Infrared Analysis Third Edition* (New York: CRC Press)

[2] Reynertson AK and Mahmood K 2015 *Botanicals Methods and Techniques for Quality and Authenticity* (New Jersey: CRC Press)

[3] Njintang NY, Mbofung CMF and Kesteloot R 2007 *J. Food Eng.* **81** 250-6

[4] Miller JN and Miller JC 2010 *Statistics and Chemometrics for Analytical Chemistry 6th Edition* (Harlow: Pearson Education)

[5] Lebot V, Malapa R, and Bourrieau M 2011 *J. Agric. Food Chem.* **59** 9327-9334

[6] Sanchez MT, Varo AG, Guerrero JE, and Marin DP 2013 *Postharvest Biol. Technol.* **85** 116-23.

[7] Bertone E, Venturello A, Giraudo A, Pellegrino G, and Geobaldo F 2016. *Food Control* **59** 683-9.

[8] Miranda J R, Lopez I R, Lara E H, Sanchez M, Licon E D and Vera V 2011 *LWT-Food Sci Tech.* **44** 673-80

[9] Gad H A, El-Ahmady S H, Abou-Shoer M I and Al-Azizi M M 2012 *Phytochem. Anal.* **24** 1-24

# Classification of java tea (*Orthosiphon aristatus*) quality using FTIR spectroscopy and chemometrics

**R Heryanto[1,2*], D I Pradono[1,2], E Marlina[2] and L K Darusman[1,2]**
[1]Biopharmaca Research Center, Bogor Agricultural University, IPB Taman Kencana Campus, Jl. Taman Kencana No 3, Bogor, Indonesia.
[2]Department of Chemistry, Faculty of Mathematics and Natural Sciences, Bogor Agricultural University, Bogor, Indonesia.

E-mail: kemometrik@gmail.com

**Abstract.** Java tea (*Orthosiphon aristatus*) is a plant that widely used as a medicinal herb in Indonesia. Its quality is varying depends on various factors, such as cultivating area, climate and harvesting time. This study aimed to investigate the effectiveness of FTIR spectroscopy coupled with chemometrics for discriminating the quality of java tea from different cultivating area. FTIR spectra of ethanolic extracts were collected from five different regions of origin of java tea. Prior to chemometrics evaluation, spectra were pre-processed by using baselining, normalization and derivatization. Principal Components Analysis (PCA) was used to reduce the spectra to two PCs, which explained 73% of the total variance. Score plot of two PCs showed groupings of the samples according to their regions of origin. Furthermore, Partial Least Squares-Discriminant Analysis (PLSDA) was applied to the pre-processed data. The approach produced an external validation success rate of 100%. This study shows that FTIR analysis and chemometrics has discriminatory power to classify java tea based on its quality related to the region of origin.

## 1. Introduction

The use of java tea as a medicinal herb makes it a good potential in pharmaceutical world. Currently, people choose herbal remedies because they are rarely causing side effects and the price is lower than synthetic medicines. However, there are problems in the production of herbal medicines, such as availability and quality of raw materials, materials standardization, quality stability and control that are considered as challenging [1]. Bioactive components contents of medicinal herbs vary widely depends on the species, variety, geographical origin, cultivation, harvesting method, and post-harvest processing. These variations may cause inconsistency in the efficacy, quality, and safety of the herbal products. Therefore, it is needed a serious control in the determination of specifications and parameters for raw materials [2].

Qualitative method frequently used to analyze active compounds is FTIR (fourier transform infrared) spectroscopy providing information about the presence of functional groups in a sample and commonly used to see the fingerprint of a sample. Every type of active compound will give a unique IR spectrum that

* Corresponding author

may be used to see the response consistency of raw material quality standardization. Other methods generally used in the quality control of herbal medicines raw materials are HPLC, GC, GC-MS, and TLC. These methods, however, need sample preparation and a relatively long time of examination process.

Medicinal herb analysis using FTIR spectrophotometer generates a very complicated FTIR spectra since this spectra is a result of chemical compounds interaction in a complex sample matrix. This spectra is difficult to be directly interpreted that it is needed chemometrics, a method used to get both qualitave and quantitave information from the spectra. Combination of infrared spectrometry and chemometrics has been widely used as a rapid quality control method of herbal medicines with extensive variety and for example, it has been used for discrimination of *Gentiana rigescens* from different origins [3].

Multivariate analysis frequently used are PCA, PLSDA, and PLS, among others. PCA and PLSDA may be used to group samples, while PLSDA is a common regression technique for multivariate data that may be used to predict hidden information in mixed spectra [4].

This study aimed to conduct quality control of java tea raw materials from several regions having relatively different geographical conditions by grouping and making a prediction model to relate the IR spectra characteristics and java tea raw materials. The parameters of the raw materials and extracts tested are ash content, water content, total phenol, and total flavonoid.

## 2. Experimental Section

This study consisted of 3 main working steps, sample preparation, quality analysis, and chemometrically data processing. Sample preparation includes harvesting, drying, and sample extraction. Quality analysis was conducted to dried samples and to the obtained extracts in the sample preparation step. Dried samples were tested for water and ash contents, extract samples were analyzed for total phenol and total flavonoid contents. FTIR analysis was also performed to java tea extracts to obtain IR spectra. Absorbance data and the wavelength of the IR spectra were chemometrically processed using The UnscramblerX 10.3. Chemometric methods used were PCA and PLSDA. The results from the chemometrically processed data and quality data from the quality analysis generated a model of java tea classification from 5 different regions.

### *2.1 Sample Preparation [5,6]*

The leaves of white-flowered java tea from 5 different regions were harvested in the afternoon. The leaves were dried using oven at 50°C for ±24 hours. Powdered samples were prepared and sieved using 40-meshed sieve. Ten gram of sample was then extracted in 5 replicates for each region using 100 mL ethanol 40%. The extract was then put in a water bath shaker for 120 minutes at 65°C. Subsequently, the resulting filtrate was filtered by using filter paper. This extract was concentrated with rotary evaporator and then kept in a bottle and put in refrigerator.

### *2.2 Total Ash Assay [7]*

Two gram of sample was put in a pre-combusted and pre-weighted silicate crucible. The tested sample was slowly combusted until the charcoal was no longer left, ashed at 600°C until complete ashing was achieved, then cooled and weighted. Combustion was repeated until it was obtained a constant weight. Total ash content was calculated as %w/w of sample.

### *2.3 Water Content Assay [8]*

Empty dish was dried in an oven at 105°C for 3 hours, transfer to desiccator to cool, and weighted. Two gram of sampel (a) was weighted and put in the dish, and dried at 105°C for 3 hours, transfer to desiccator to cool and weighted. The dish and the sample were reweighted until it was obtained a constant weight (b).

$$\text{Water Content (\%)} = \frac{a-b}{a} \times 100\%$$

*2.4 Total Phenol Assay [9]*
Total phenol content in an extract was determined by using Follin-Ciocalteu (FC) reagent and external calibration curve was created by using gallic acid. 10 mg Extract was weighted, dissolved in 5 mL of water, and diluted 20X. To 3 mL aquadest, 2 mL absolute ethanol, and 0.5 mL 50% FC reagent (v/v) it was added 2 mL extract. The mixture was left for 5 minutes and then 1 mL 5% $Na_2CO_3$ (w/v) was added. The mixture was homogenized and then incubated in the dark for 1 hour. The absorbance was measured with UV-VIS spectrophotometer at 725 nm.

*2.5 Flavonoid Content Assay [7]*
Two hundred milligram of extract was hydrolized with 1 mL 0.5% hexamethyltetraamine 0.5% (w/v), 20 mL aceton, and 2 mL HCl 25% in water. Hydrolisis was performed by refluxing for 30 minutes. The hydrolized mixture was filtered with cotton into a 100 mL volumetric flask and aceton was added up to the measuring mark. 20 mL Filtrate was then transferred into a separatory funnel, added with 20 mL aquadest, and extracted 3 times, each with 15 mL ethyl acetate. Ethyl acetate fractions were collected into a 50 mL volumetric flask and added up with ethyl acetate up to the measuring mark. Spectrophotometry analysis was started by transferring 10 mL of the ethyl acetate fraction solution into a 25 mL volumetric flask and 1 mL 2% $AlCl_3$ in 5% glacial acetic acid (in methanol) was added. 5% Glacial acetic acid was added up to the measuring mark and the absorbance was measured at 425 nm. Pure quercetin solution in acetic acid was used as standard solution.

*2.6 FTIR Analysis*
In order to make a pellet, 2 mg of ethanol extract was mixed with 200 mg KBr. The pellet was made by using a hand press with pressure of 80 kN for 10 minutes. The spectrum was measured using FTIR spectrophotometer in the range of 4000-400 cm-1. The spectrum data was normalized that the lowest data value was set to 0, while the highest value was set to 2. The result of normalization was baselining-corrected to make the spectra baseline in the 0 absorbance, followed by first derivatization and Savitsky Golay method smoothing.

*2.7 Chemometrically Data Analysis*
FTIR spectra were saved in OPUS format containing data in the absorbance mode. There were total thirty five samples spectra where seven samples came from each origin. Twenty five spectra were used as training data and two spectra from each origin were used for test data. Samples grouping were performed by using PCA and PLSDA. PCA and PLSDA analysis of the resulting FTIR spectra used The UnscramblerX 10.3 software.

*2.8 Statistical Data Analysis*
The mean values of the data from water content, ash content, extract yield, total phenol and total flavonoid were determined and standard deviations were also determined. Further analysis was performed for variance and Duncan's Multiple Range Test.

## 3. Results and Discussion

*3.1 In Water Content, Ash Content, and Extract Yield*
Water and ash content assays in this study were performed by gravimetric methods. The water and ash contents of dried samples from the five regions had fulfilled the maximum requirements in FHI (2008),

which are less then 10%. Water content shows the water contained in the sample materials. Water content assay may assist in the determination of the material actual weight and used in the extract yield calculation. The lower the water content, the higher the material stability and the lower the material destruction [10]. On the other hand, ash content shows the internal and external mineral contained in the material and related to the purity of the materials as well as materials contamination [11]. The values of water content, ash content, and extract yield are significantly different ($p<0.05$) for some particular regions (Figure 1).

The lower the ash content, the lower the possibility of the material to be contaminated. The heavier the extract yield, the more chemical compounds extracted from the raw materials. In general, quality is said to be good if the water content is low, the ash content is low, and the yield is high. Samples with the lowest water content comes from the sample from Nagrak with value significantly different from the samples from the other regions. The lowest water content value from the samples from Cigombong is indistinguishable with the samples from Leuwiliang and has the highest extract yield compared to the other regions (Figure 1).

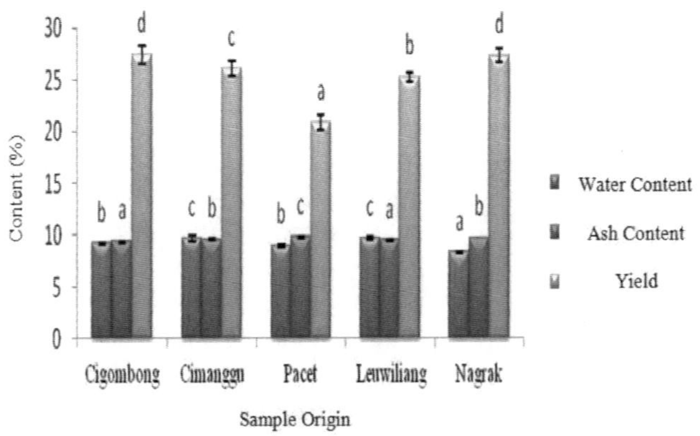


**Figure 1.** Water content, ash content and extract yield. Values are given in mean ± sd. Values marked with (a-d) are significantly different ($p<0.05$)

*3.2 Metabolites Content and Antibacterial Activity of Java Tea Leaves Extract*

Analysis of the chemical compounds content in java tea leaves extracts was performed to the total phenol and total flavonoid contents. In java tea, phenol is an important compound since it has a significant role as an antioxidant. There are 20 types of phenol in java tea: 9 lipophilic flavons, 2 flavonol glycosides, and 9 caffeic acid derivatives [12]. Flavonoid is a main group of phenol in plants and sinensetin belongs to the group of methoxy flavon or lipophilic flavonoid [13].

The samples from Nagrak, based on the results of this study, has the highest total phenol content and significantly different from the samples from the other regions (Figure 2). The samples from Cimanggu has the highest total flavonoid content but insignificantly different with the samples from Nagrak. In general, the samples from Nagrak and Cimanggu have higher chemical content quality compared to other samples. On the contrary, the samples from Pacet has the lowest quality.

The results of chemical compounds content analysis correspond with the geographical characteristics of respective regions. Java tea grows very well in the altitude of 100-1000 mdpl, tropical climate, average rain fall of 3000 mm/year, and full sunlight [14]. Nagrak and Cimanggu meet the requirements for the growth of java tea. Cigombong has lower rain fall, Leuwiliang has lower temperature, while Pacet either has lower temperature or height that achieve 1400 mdpl. Further, the planting pattern of Pacet and Leuwiliang is polyculture, while the others monoculture. According to [10], other factors that may influence the quality of herbal medicines such as species variation, harvesting time, the part of the plant used, and post-harvest treatments are considered as do not give any effects because, in this study, they are same for each region.

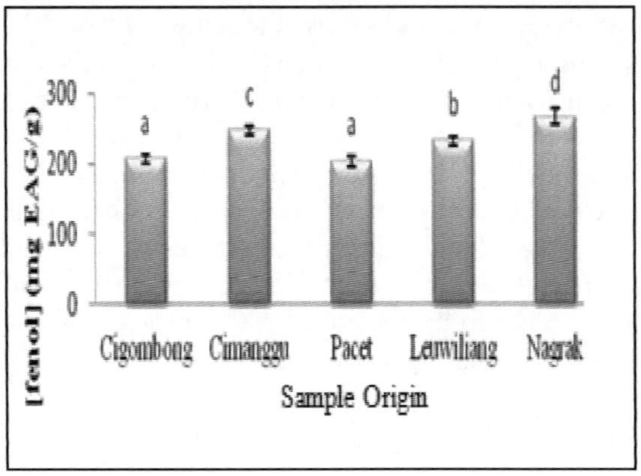


(I)

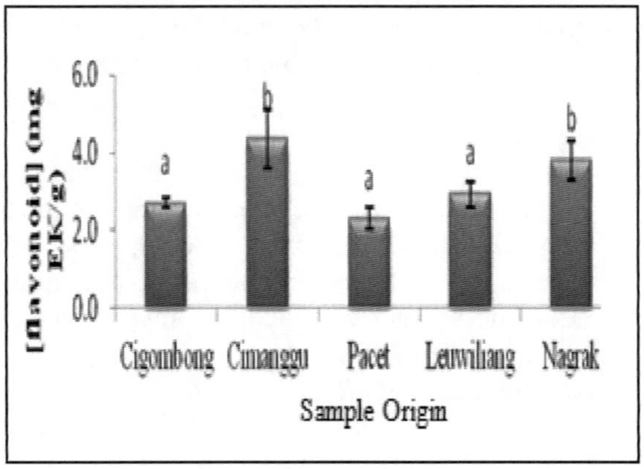


(II)

**Figure 2.** Total Phenol (I) and Total Flavonoid (II) contents of sample extracts from 5 regions. The values are given in mean ± sd. The values marked with (a-d) are significantly different ($p<0.05$)

Linear correlation between total phenol content and total flavonoid content in the sample can be seen from the results of Pearson Correlation test. The coefficient of correlation between total phenol and total flavonoid is 0.744, showing that the total phenol and total flavonoid contents are positively correlated and this correlation is relatively strong. Total phenol contents in the samples are high and so are the total flavonoid contents (Table 1).

**Table 1.** Pearson Correlation Analysis

| | Total Phenol | Total Flavonoid |
|---|---|---|
| Total Phenol | 1 | |
| Total flavonoid | 0.744 | 1 |

The correlation value between phenol and flavonoid is relatively strong, which is due to most flavonoids belong to phenol. However, there are also caffeic acid derivatives, tannins, and other compounds in the samples that are not belong to phenol.

*3.3 FTIR Spectra of Java Tea Leaves Extracts*

Every compound in a medicinal herb has an important role in a complicated mixed system for its effect to the efficacy of the plant. FTIR spectra generated (Figure 3) are the absorbances of various chemical components in the java tea leaves extract. The FTIR spectra of the samples from the five regions are not visually different showing that the chemical components contained are generally the same.

A strong and wide absorbance observed at the wave number of about 3400 $cm^{-1}$ shows the presence of hydroxyl group like in polyphenol, and at 2962 $cm^{-1}$ and 2872 $cm^{-1}$ show C-H of aldehyde, at 1600 $cm^{-1}$-1700 $cm^{-1}$ show C=O, 1600-1420 $cm^{-1}$ shows phenyl, 1456 $cm^{-1}$ and 1382 $cm^{-1}$ show C-H of $CH_3$, and 1270-1150 $cm^{-1}$ shows ester (C-O) [15].

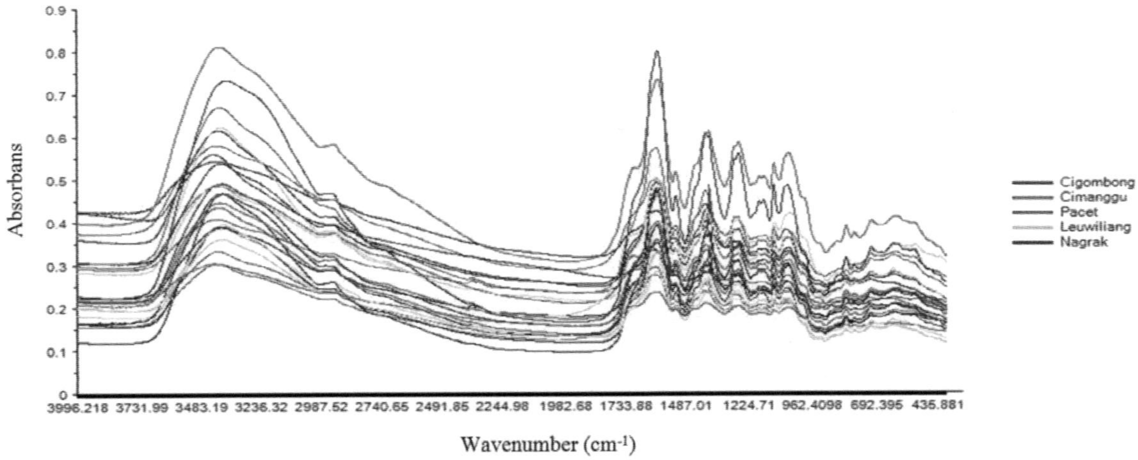


**Figure 3.** . FTIR Spectra from 25 samples of 5 different regions

### *3.4 FTIR Spectra Grouping using PCA and PLSDA*

FTIR spectra groupings of java tea leaves extracts from the five regions were performed by PCA. This method is not able to group the initial sample spectra or spectra without any pre-processes (Figure 4a). This is because the initial spectra are still influenced by baseline shifts, difference in the number of samples analyzed, and the noise produced by the detector. This influence may be resolved by pre-processing the spectra including baseline correction, normalization, and derivatization. These pre-processing techniques could improve PCA capability to group samples without loosing much information with 73% total variance obtained.

Figure 4b shows samples grouping from the five regions and each region was marked with different color. Samples from Pacet (P) found in quadrant I, samples from Cigombong (G) found in quadrant II and III, samples from Leuwiliang (L) and Cimanggu (C) found between quadrant I and IV, while samples from Nagrak (N) found in quadrant IV. Samples from Cigombong exhibits a very clear separation from the samples of the other four groups of regions. This could mean that there are possibilities the Cigombong samples have quality characteristics and chemical contents different from the samples of the other regions.

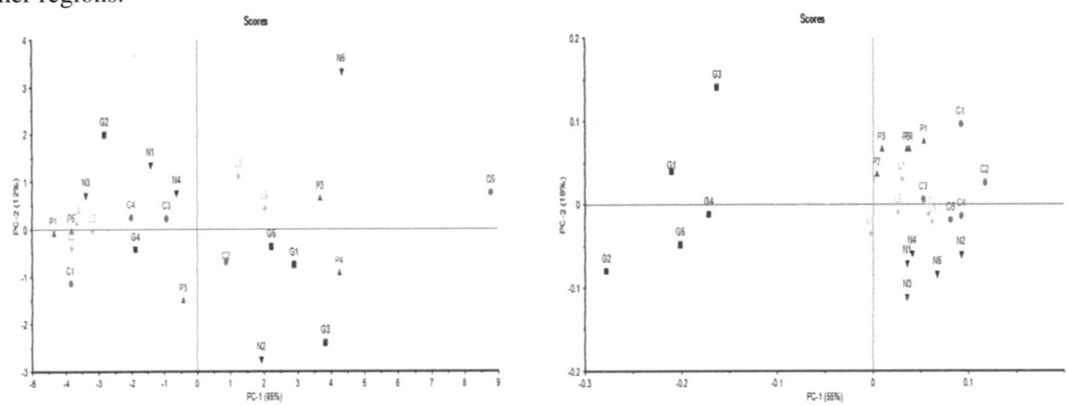

(a) (b)

**Figure 4.** PCA Score plot (a) spectra without pre-processing (b) spectra with pre-processing

Quality groupings in Table 2 aim to facillitate the quality evaluation of the samples from each region. The quality group I is a group of samples with the best criteria compared to other samples based on significance test and the quality tested. The best quality comes from the samples from Nagrak because all test results are in the quality group I, except for ash content. Samples from Cimanggu is in the second place after Nagrak, and the lowest is the samples from Pacet. These results correspond with the results of sample grouping using PCA. The sample with the best quality lies in quadrant IV, medium quality lies between quadrant I and IV, and the lowest quality lies in quadrant I. Samples from Cigombong has good physical quality, but low in chemical content quality.

**Table 2.** Quality evaluation of extracts from each region

| Group of Quality | Quality | | | | |
|---|---|---|---|---|---|
| | Water Content | Ash Content | Extract Yield | Total Phenol | Total Flavonoid |
| I | N | G,L | G,N | N | N,C |
| II | G,P | C,N | C | C | G,P,L |
| III | C,L | P | L | L | |
| IV | | | P | G,P | |

Samples grouping by using PLSDA was performed to 2 matrix, i.e. the absorbance data from the results of FTIR analysis as X matrix and response matrix for every sample's region as Y matrix. Response value of 1 is given for the samples successfully predicted to their own region of origin and 0 for the samples predicted not to their own region of origin. Calibration model was well built for having R2 approaching 1 and RMSE (Root Mean Square Error) approaching 0 for all samples from each region. Prediction models of samples from Cimanggu and Pacet are relatively good compared to the other 3 regions (Table 3).

**Table 3.** Criteria of the Goodness of fit from PLSDA Model

| Sample | Callibration | | Prediction | |
|---|---|---|---|---|
| | $R^2$ | RMSEC | $R^2$ | RMSEP |
| **G** | 0.9739 | 0.0646 | 0.9061 | 0.1313 |
| **C** | 0.9792 | 0.0577 | 0.7578 | 0.2109 |
| **P** | 0.9948 | 0.0289 | 0.9304 | 0.1131 |
| **L** | 0.9789 | 0.0581 | 0.8762 | 0.1508 |
| **N** | 0.9790 | 0.0579 | 0.8561 | 0.1626 |

The results of prediction of 2 samples randomly selected from each region can be seen in Table 4. The prediction values will approach 1 if the samples from one region tested to the model of its own region and vice versa. The results of prediction are not sufficiently good for the samples from Leuwiliang and Nagrak, replicate 1. This may be due to the low prediction capability of the model and difference in the FTIR spectra characteristics of the samples tested.

**Table 4.** Sample prediction using PLSDA model

| Sample | Replicate | Prediction Value in PLSDA Model | | | | |
|---|---|---|---|---|---|---|
| | | Cigombong | Cimanggu | Pacet | Leuwiliang | Nagrak |
| Cigombong | 1 | 0.9029 | 0.3485 | -0.0291 | -0.1930 | -0.0292 |
| | 2 | 0.7750 | 0.1535 | 0.0521 | -0.0327 | 0.0521 |
| Cimanggu | 1 | -0.0290 | 1.2399 | -0.0320 | -0.1168 | -0.0621 |
| | 2 | 0.0853 | 0.7219 | -0.1987 | 0.1798 | 0.2117 |
| Pacet | 1 | 0.0319 | -0.4598 | 0.7702 | 0.2986 | 0.3590 |
| | 2 | 0.0395 | 0.1823 | 1.0255 | -0.2018 | -0.0455 |
| Leuwiliang | 1 | 0.1085 | 0.2847 | 0.0624 | 0.5712 | -0.0268 |
| | 2 | -0.0823 | 0.0820 | -0.0722 | 1.0709 | 0.0016 |
| Nagrak | 1 | -0.1232 | 0.3936 | 0.1720 | -0.1053 | 0.6628 |
| | 2 | 0.0446 | 0.1379 | -0.0059 | -0.0807 | 0.9041 |

## 4. Conclusion

Extract samples of java tea leaves from 5 different regions are succesfully grouped based on the results of PCA analysis and the physicochemical quality. The samples from Nagrak have better quality than the other regions, in which the score plot lies in quadrant IV. Medium quality lies in quadrant IV and I, which is for the samples from Cimanggu and Leuwiliang. The quality for the samples from Pacet is not sufficiently good and lies in quadrant I. The samples from Cigombong tend to be different from other samples, which lies between quadrant II and III and have low total phenol and total flavonoid contents. Samples grouping was also performed using PLSDA and almost all samples could be predicted based on their respective region of origin.

## References

[1] Bandaranayake WM 2006 *Modern Phytomedicine. Turning Medicinal Plants into Drugs.* (Weinheim: WILEY-VCH Verlag Gmbh & Co. KGaA)

[2] Komarawinata HD 2008 *Budidaya dan pasca panen tanaman obat untuk meningkatkan kadar bahan aktif.* (Bandung: Unit Riset dan Pengembangan, PT Kimia Farma (Persero) Tbk.)

[3] Zhao Y, Zhang J, Jin H, Zhang J, Shen T, Wang Y 2015 *J. AOAC Int.* **98** 22-6

[4] Brereton RG 2003 *Chemometric: Data Analysis for the Laboratory and Chemical Plant* (England: John Wiley & Sons. Ltd.)

[5] Abdullah S, Ahmad MS, Shaari AR, Johar HM, and Noor NFM 2011 *Int. Conference on Food Engineering and Biotechnology* vol 9 p 305-9.

[6] Chew KK, Khoo MZ, Ng SY, Thoo YY, Wan Aida, WM, and Ho CW 2011 *Int. Food Res. J.* **18** 1427-35.

[7] [Depkes] Departemen Kesehatan 2008 *Farmakope Herbal Indonesia* ed 1 Jakarta (ID): Depkes

[8] [AOAC] Association of Official Analytical Chemistry 2005 *Official Method of Analysis* (Arlington: The Association of Official Analytical Chemist. Inc.)

[9] Santoso J, Anwariyah S, Rumiantin RO, Putri AP, Ukhty N, and Stark YY 2012 *J. Coastal Dev.* **15** 189-96

[10] Kunle OF, Egharevba HO, and Ahmadu PO 2012 *Int. J. Biodivers. Conserv.* **4** 101-12

[11] Emilan T, Kurnia A, Utami B, Diyani LN, and Maulana A 2011 *Konsep Herbal Indonesia: Pemastian Mutu Produk Herbal* [bibliografi] (Depok: Universitas Indonesia)

[12] Akowuah GA, Zhari I, Norhayati I, Sadikun A, and Khamsah SM 2004 *Food Chem.* **87** 559-66

[13] Hossain MA and Ismail Z 2012 *Arab J. Chem.* **2** 1–4

[14] Sembiring BS, Rizal M, and Suhirman S 2012 *Budidaya dan Pascapanen Kumis Kucing (Orthosiphon stamineus Benth)* (Bogor: Balai Penelitian dan Pengembangan Pertanian, Pusat Penelitian dan Perkebunan, Balittro)
[15] Pavia DL, Lampman GM, and Kriz GZ 2001 *Introduction to Spectroscopy* (Washington (US): Thomson Learning Inc.)

**Acknowledgement**
This research was carried out with the aid of a research grant from SINAS of Ministry of Research, Technology and Higher Education.

IOP Conf. Series: Journal of Physics: Conf. Series **835** (2017) 012013 doi:10.1088/1742-6596/835/1/012013

# Overproduction, purification and characterization of human interferon alpha2a-human serum albumin fusion protein produced in methilotropic yeast *Pichia pastoris*

**R A Ningrum*, A Santoso and N Herawati**

Research Center for Biotechnology, Cibinong Science Center,
Indonesian Institute of Sciences
Jalan Raya Bogor km 46, Cibinong, Bogor, West Java, Indonesia, 16911

E-mail : rati004@lipi.go.id / ratih.asmana@gmail.com

**Abstract.** Human interferon alpha2a (hIFNα2a) is a therapeutic protein that used in cancer and hepatitis B/C therapy. The main problem of using hIFNα-2a is its short elimination half life due to its low molecular weight. Development of higher molecular weight protein by albumin fusion technology is a rational strategy to solve the problem. In our previous research we constructed an open reading frame (ORF) encoding hIFNα2a-human serum albumin (HSA) fusion protein that expressed in *Pichia pastoris (P. pastoris)* protease deficient strain SMD1168. This research was performed to overproduce, purify and characterize the fusion protein. To overproduce the protein, cultivation was performed in buffered complex medium containing glyserol (BMGY) for 24 h and protein overproduction was applied in buffered complex medium containing methanol (BMMY) for 48 hours at 30°C. The fusion protein was purified by blue sepharose affinity chromatography. Molecular weight characterization by SDS PAGE corresponds with its theoretical size, 85 kDa. Western blot analysis demonstrated that the fusion protein was recognized by anti hIFNα2 and anti HSA monoclonal antibody as well. Amino acid sequence of the fusion protein was determined by LC MS/MS2 mass spectrometry with trypsin as proteolitic enzyme. There were three fragments that identified as hIFNα2a and seven fragments that identified as HSA. Total identified amino acids were 150 residues with 20% coverage from total residues. To conclude, hIFNα2a-HSA fusion protein was overproduced, purified and characterized. Characterization based on molecular weight, antibody recognition and amino acid sequence confirmed that the fusion protein has correct identity as theoretically thought.

## 1. Introduction

hIFNα2a is a cytokine that consists of 165 amino acids with 19 kDa in size. It has wide biological activity such as antivirus, antiproliferation as well as immunomodulation [1]. Recombinant hIFNα2a has already used as therapeutic protein in hepatitis B/C and several cancer treatments since 1986 [2]. The major problem of using hIFNα2a for patient treatment is short elimination half life due to rapid renal clearance. As low molecular weight protein, major elimination route of the protein in the body is through renal filtration system by glomerulus. The consequence of short half-life time is high

* Corresponding author

frequency treatment to maintain the effective concentration of the protein. The treatment frequency for hepatitis B or C is three times per week for 24 to 48 weeks and for oncological indication is daily from several months to a year or longer [3].

The strategies to improve the elimination half-life time of rhIFNα2a have been performed with several approaches. The most successful is higher molecular weight modification by pegylation. The commercial pegylated protein PEGASYS (Hoffman La Roche, UK) with 40 kDa PEG molecule has 70 hour of elimination half life that decreases the therapy frequency become only once a week [4]. Albumin-fusion technology is a rational, simple and flexible alternative strategy to increase the protein molecular weight through a fusion between rhIFNα2a and human serum albumin (HSA). HSA has a 9 day long half-life, widely distributed in the body and not immunogenic. Moreover it may improve the solubility and stability of the protein [5].

The albumin fusion technology widely used as fusion partner to extend in vivo half life of protein therapeutics at its N terminal or C terminal. Zhao et al. reported that HSA fusion at the N-terminal of rhIFNα2b caused instability due to the disruption of disulfide bond that formed by cysteine at positions 1 to 98. The heterogeneity can be avoided by adding a linker between the two proteins or altering the orientation of HSA fusion at the N-terminal into C-terminal [6]. The study of rhIFNα2b-HSA fusion protein reported the improvement of elimination half-life from 5 hours into 159 hours that decrease therapy frequency become once in two week [5]. In our previous research we synthesized an ORF encoding rhIFNα2a-HSA fusion protein. The constructed ORF in frame with α-factor signal sequence to obtain extracellular fusion protein. This research was purposed to overproduce, purify and characterize the rhIFNα2a-HSA fusion protein.

## 2. Materials and Methods

### 2.1 Overproduction

Overproduction was performed using shake flask method with 50 mL of BMMY media. Single colony was grown overnight in 25 ml BMGY medium at 30°C and 250 rpm until log phase ($OD_{600}$= 2-6). The culture was centrifuged at 1500 x g for 5 min. The pellet was resuspended in 50 ml BMMY medium ($OD_{600}$ = 1.0 containing 1.5% methanol as inducer). Protease inhibitor cocktail (Roche) was added to overcome protein degradation. The 48h harvesting time was carried out by repeating the induction after 24 h cultivation time. Harvesting was performed by centrifugation at 1500 x g for 5 minutes at room temperature to collect supernatant.

### 2.2 Purification

Supernatant containing fusion protein was concentrated 10X (v/v) using tangential filtration system with 10 kDa molecular weight membrane cut off. Concentrated supernatant was purified using affinity chromatography method by blue sepharose. The concentrated fraction was applied into prewashed blue sepharose resin (GE Healthcare). The prewashed was applied using PBS 1x pH 7.0 as wash buffer. After 3 times sample reloaded, the resin was washed 5 times using wash buffer, then eluted with 5 mL elution buffer containing PBS 1x pH 7.0 and 1.5 M NaCl. The purified protein was filtered by Amicon® (Millipore) with 10kDa molecular weight cut off.

### 2.3 Characterization

Protein molecular weight was characterized by SDS-PAGE method with 10% (w/v) of polyacrylamide gel and coomassie blue staining solution (BioRad). Western blot used to identify the protein based on specific monoclonal antibody recognition. The primary antibodies was mouse anti IFNα2 (Merck 407290-500 UGCN, Germany) with 1: 1000 dilution or mouse anti HSA (Sigma, USA) with 1:1000 dilution. The secondary antibody was 1: 7500 dilution of anti mouseIgG alkaline phosphatase conjugate (Promega, USA) with NBT/BCIP detection (Merck, Germany). Amino acid sequence was determined by LC MS/MS2 method (Proteomic Inc). The samples prepared from coomassie blue stained gel slice.

## 3. Result and Discussion

The overproduction of fusion protein in the *P. pastoris* consisted of biomass increment in BMGY media and protein production in BMMY media. Glycerol in the BMGY media is the main carbon source and repressor of alcohol oxidase 1 (*AOX1*) promoter. BMMY media provides methanol as carbon source [7]. The concentration of methanol in the medium determined the efficiency of the recombinant protein production. Overproduction in *P. pastoris* depends on biosynthesis of alcohol oxidase (AOX) which is regulated by *AOX1* promoter. We constructed an ORF encoding fusion protein in the downstream of *AOX1* promoter, therefore its expressions would be in line with AOX. *P. pastoris* can produce proteins intracellularly and extracellularly. In this study the ORF constructed fusion protein contains α-factor preprotein signal peptides, so the fusion protein produced extracellularly.

One of disadvantages high cell density overproduction is protein instability due to protease activities. Combination of high density culture and small percentage cell lysis in the media may cause some proteins rapidly degraded. The protease such as proteinase A and proteinase B, carboxylpeptidase and aminopeptidase are major known proteases that induce protein degradation in *P. Pastoris* [8]. Several strategies to inhibit proteolysis activity have been reported, such as protein engineering [9], optimization of fermentation parameters (pH, temperature and growth rate), modification of culture media composition (rich media, additional amino acid or peptone, lowering salt concentration as well as soytone addition) and the use of protease deficient strain [10]. In this study we used protease deficient strain SMD1168 that lack of proteinase A activity. Protease inhibitor also used in overproduction media to reduce proteolysis as well. After 48h of overproduction, the supernatant containing fusion protein was collected, purified and characterized.

Fusion protein purification performed by affinity chromatography using blue sepharose (GE healthcare) resin that consists cross-linked 6% agarose beads modified with Cibacron Blue 3G covalently attached by the triazine coupling method. The blue dye binds many proteins including albumin and interferon. Molecular weight characterization of purified fusion protein informed that the protein was 85 kDa in size which corresponds with its theoretical size (Figure 1). Further characterization using anti HSA and anti hIFNα2 monoclonal antibodies showed that the protein was recognized by both antibodies (Figure 2). This result strongly indicated the correct identity of fusion protein. Amino acid sequence confirmed the antibody based characterization with ten identified peptides. By using trypsin as proteolitic enzyme, there were three fragments that identified as hIFNα2a and seven fragments that identified as HSA. Total identified amino acid were 150 residues with 20% coverage from total residues (Figure 3).

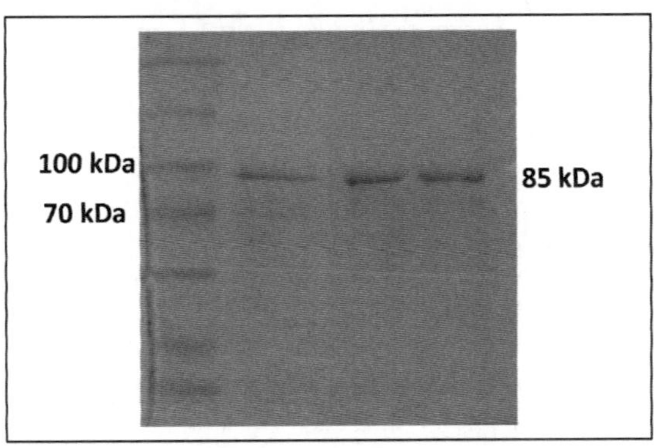


**Figure 1.** Molecular weight analysis of fusion protein by SDS PAGE, 1=eluate 1; 2=eluate 2; 3= eluate 3.

IOP Conf. Series: Journal of Physics: Conf. Series **835** (2017) 012013 doi:10.1088/1742-6596/835/1/012013

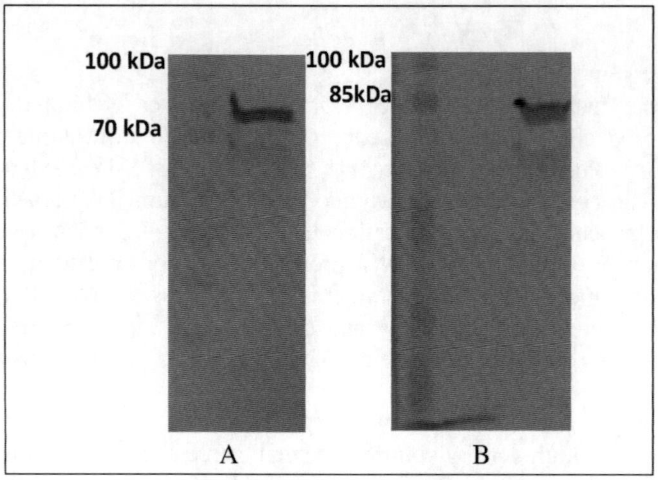


**Figure 2.** Monoclonal antibody recognition-based analysis of fusion protein, A= using anti hIFNα2 antibody, B=using anti HSA antibody

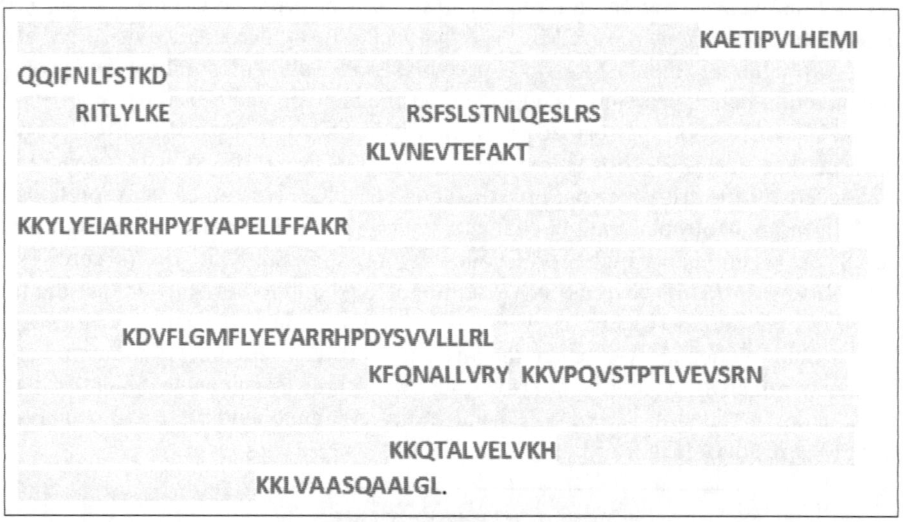


**Figure 3.** Amino acid sequence determination of fusion protein. hIFNα2a amino acids shown in blue and HSA amino acids shown in red.

## 4. Conclusions

Fusion protein has been overproduced, purified as well as characterized. The identified fusion protein has 85 kDa molecular weight in size and confirmed as fusion protein between hIFNα2a and HSA.

## References

[1] Ningrum R A 2014, *Scientifica* **2014** 1-8
[2] Jonasch E and Haluska FG, 2001 *Oncologist* **6** 34-55
[3] Ceaglio N, Etcheverrigaray M, Kratje R and Oggero M 2008 *Biochimie* **90** 437-49
[4] Reddy R, Modi M W and Pedder S 2002 *Adv. Drug. Deliv. Rev.* **54** 571–86
[5] Subramanian M G, Fiscella M, Smith A L 2007 *Nat. Biotechnol.* **25** 1411-9
[6] Zhao H L, Yai X Q, Xue C, Wang Y, Xiong X H and Liu Z M 2008 *Protein Expr. Purif.* **6** 73-7
[7] Cregg J M, Lin C J, Shi J and Higgins D R 2000 *Mol. Biotechnol.* **16** 23-5.
[8] Sinha J, Plantz B A, Inan M and Meagher M M 2005 *Biotechnol. Bioeng.* **89** 102–12

[9] Gustavsson M, Lehtio J and Denman S 2001 *Protein Engineering* **14** 711–5
[10] Zhao H L, Yai X Q, Xue C, Wang Y, Xiong X H and Liu Z M 2012 *Mol. Pharm.* **9** 664-70

**Acknowledgement**
This research was not possible completed without the support of Riset Unggulan LIPI 2016 from Indonesian Institute of Sciences.

IOP Conf. Series: Journal of Physics: Conf. Series **835** (2017) 012014 doi:10.1088/1742-6596/835/1/012014

# Thin layer chromatography fingerprint, antioxidant, and antibacterial activities of rhizomes, stems, and leaves of *Curcuma aeruginosa* Roxb.

**A Safitri[1], I Batubara[1,2*] and N Khumaida[3]**

[1]Department of Chemistry, Faculty of Mathematics and Natural Sciences, Bogor Agricultural University, Jalan Agatis Kampus IPB Dramaga, Bogor 16680, Indonesia
[2]Tropical Biopharmaca Research Center, Bogor Agricultural University, Jalan Taman Kencana No 3 Bogor, 16128, Indonesia
[3]Department of Agronomy and Horticulture, Faculty of Agricultural, Bogor Agricultural University, Jalan Meranti Kampus IPB Dramaga, Bogor 16680, Indonesia

E-mail: ime@apps.ipb.ac.id

**Abstract.** Fingerprints of 5 *temu hitam* (*Curcuma aeruginosa* Roxb.*)* accessions (Malang, Cirebon, Kuningan 1, Bogor, and Liwa) were determined by thin-layer chromatography (TLC) and compared to fingerprints of turmeric (*Curcuma longa* L), *temu putih* (*Curcuma zedoaria* (Christm.) Roscoe), and *temu lawak* (*Curcuma zanthorriza* Roxb.). Maceration method with ethanol as the solvent was used for extraction. The eluent used for fingerprint by TLC was chloroform:dichloromethane (9:1v/v). Five accessions of *temu hitam* show similar fingerprint patterns, but different in band thickness. *Temu hitam* rhizomes have bands of curcuminoid (Rf 0.22, 0.10, 0.03), and characteristic bands of Rf 0.42, 0.27, and 0.77, which can be distinguished from turmeric and *temu lawak* and Rf 0.13, which is different from *temu putih*. Leaves and stems of *temu hitam* can be distinguished from *temu putih*, turmeric, and *temu lawak* at Rf 0.60. Rhizomes of all plants reveal strong antibacterial activity against *Staphylococcus aureus* and antioxidant activity on DPPH radicals than its corresponding stems and leaves. Antibacterial and antioxidant activities were determined by microdilution and TLC-bioautography. Antibacterial activity of rhizomes of Cirebon and Kuningan 1 accessions are higher than that of other accessions (MIC= 250 μg/mL; MBC= 500 μg/mL, but lower as compared to that of *temu lawak* (MIC= 62.5 μg/mL, MBC= 250 μg/mL) and tetracycline (MIC=MBC= 15.63 μg/mL). Rhizome of Liwa accession exhibits the highest antioxidant activity ($IC_{50}$= 124.88 μg/mL) amongst all accessions, but lower than that of *temu lawak* ($IC_{50}$= 18.45 μg/mL), turmeric ($IC_{50}$= 18.82 μg/mL), and *temu putih* ($IC_{50}$= 94.35 μg/mL).

## 1. Introduction

*Temu hitam* (*Curcuma aeruginosa* Roxb.) is one type of curcuma species of Zingiberaceae family. This plant can be categorized as a medicinal plant because extracts and essential oils of *temu hitam* reported active as antibacterial, antiinflammation, antipyretic, and antinociceptive agents. The ethyl acetate extract of *temu hitam* can inhibit the growth of *Pseudomonas aeruginosa*, *Staphylococcus*

* Corresponding author

*aureus*, and *Bacillus subtilis* [1] and its chloroform extract has bioactive compounds that act as antinociceptive, antiinflammation, and antipyretic [2,3]. Essential oils of *temuhitam*'s rhizomes also have antibacterial activity against *Escherichia coli*, *Bacillus aureus*, *Staphylococcus aureus*, and *Pseudomonas aeruginosa* [4]. In addition, essential oils of rhizomes, stems, and leaves of *temu hitam* can inhibit the growth of *Streptococcus mutans* and as biofilm degradator on teeth [5]. Besides those activities, ethanol and methanol extracts and essential oils of *temu hitam*'s rhizomes also have antioxidant activity [6-8].

Various activities of *temu hitam* is related to its secondary metabolites. These metabolites are formed when the plants are under stress conditions, for self-protection, and defense [9-11]. *Temu hitam* was reported to contain secondary metabolites such as terpenoids, flavonoids, tannins, alkaloids, and saponins [7,12]. The content of secondary metabolites in various plants can be different depend on the diversity of plant physiological (organ development, the type of plant materials, seasonal variation, and wound mechanical or chemical), environmental conditions (climate, pollution, disease, and edaphic factor), geographic variation, genetics and evolution, storage, and the amount of plant material [13]. The number of compounds in any kind of plant drugs affected these factors need to be evaluated, compared, and quality controlled through fingerprint analysis. Fingerprint patterns indicate the overall profile of components because it can represent the diversity of components in medicinal plants [14]).

Fingerprint analysis of *temu hitam* from three regions (Nagrek, Cikabayan, and Tawangmangu) was performed using liquid chromatography-mass spectrometer [12]. The ethanol extract of rhizomes of *temu hitam* from three regions have chemical components with different concentrations. *Temu hitam* from Nagrek has the greatest antioxidant potency and *temu hitam* from Cikabayan has the greatest toxicity. In this study, the profile of rhizomes, stems, and leaves of *temu hitam* with different accessions determined and compared with its species and other curcuma species through fingerprint analysis by thin layer chromatography. In addition, antibacterial and antioxidant activity of rhizomes, stems, and leaves of *temu hitam* and other curcuma species determined and associated with their components profile.

## 2. Methodology

### *2.1. Material*

Five accessions of fresh *temu hitam* (Malang, Kuningan 1, Liwa, Bogor, and Cirebon accession) as main samples collection of Division of Plant Biotechnology Collection, Department of Agronomy and Horticulture, IPB in Experimental Garden Sukamantri, Sukamantri Village, Tamansari Subdistrict, Bogor, West Java, and *temu lawak*, turmeric, and *temu putih* as comparator samples (Table 1) were used as plant materials. All solvents that were used are analytical grade and obtained from Merck or Sigma-Aldrich (St. Louis, USA). Beside that, TLC plate silica gel 60 $F_{254}$ (Merck, Darmstadt, Germany), standard of curcuminoid (ChormaDex, California, USA), vanillin-sulfuric acid were used for fingerprint analysis. Tryptic Soy Broth (TSB), Tryptic Soy Agar (TSA), inoculant of *Staphylococcus aureus* (IPB Culture Collection, Faculty of Mathematics and Natural Sciences, IPB), tetracycline, and DMSO 20% were used for antibacterial assay. 1,1-diphenyl-2-picrilhydrazyl (DPPH) and standard of ascorbic acid were used for antioxidant assay .

### *2.2. Procedure*

*2.2.1. Instrumentation.* Fingerprint of thin-layer chromatography was carried out using a system of semi-automatic TLC sampler Linomat V (connected to nitrogen tank), twin-through chamber, and documentation device Reprostar 3 with winCATS version 1.2.3. planar chromatography software manager (Camag, Muttenz, Switzerland). Epoch microplate Spectrophotometer with Gen5 Data Analysis software (Biotek Instruments, Inc., Winooski, VT, USA).

*2.2.2. Sample and standard preparation.* Rhizomes, stems, and leaves of each samples were dried and powdered/pulverized before extracted with ethanol. Then powder and dried plant materials were extracted with a solvent (ratio of 1 g sample: 10 ml solvent) for 24 hours three times. The extracts were filtered and concentrated in a vacum at 30-40 °C using rotary evaporator. Sample extracts were diluted with ethanol with a final concentration of 2% for fingerprint analysis. curcuminoid standard solution was prepared by dissolving curcuminoid with ethanol with a final concentration of 0.2%.

**Table 1.** Five accessions of *temu hitam* of Division of Plant Biotechnology Collection, Department of Agronomy and Horticulture, IPB.

| Accessions code of *Curcuma aeruginosa* Roxb. | Sources | Accessions code of *Curcuma aeruginosa* Roxb. |
|---|---|---|
| KUN1-JABAR | Cileuleuy Village, Cigugur Subdistrct, Kuningan, Jawa Barat | KUN1-JABAR |
| BOG-JABAR | Anyar Market, Bogor, Jawa Barat | BOG-JABAR |
| MAL-JATIM | Turirejo Village, Lawang Subdistrict, Malang, Jawa Timur | MAL-JATIM |
| LIWA-LAMP | Sebarus Village, Bukit Tinggi Subdistrict, Liwa, Lampung | LIWA-LAMP |
| CIR-JABAR | Kanoman Market, Cirebon, Jawa Barat | CIR-JABAR |
| *Curcuma longa* | Unit of conservation and cultivation of biopharmaca, Tropical Biopharmaca Research Center of IPB, Dramaga Bogor | *Curcuma longa* |

*2.2.3. Thin Layer Chromatography (TLC)-fingerprint analysis.* The ethanol extract of samples were applied to a TLC plate $G_{60}F_{254}$ with a volume of 10 μL using semiautomatic sampler CAMAG Linomat V. Then, the plates were drained and developed into a chromatography chamber with chloroform: dichloromethane (9:1v/v) as mobile phase. Temperature and humidity were not controlled, but in a controlled laboratory temperature condition. After the mobile phase reached the finish line (80 mm), plates were removed and air-dried. Furthermore, the plates were detected with UV lamp 254 nm, 366 nm, visible light, and vanillin-sulfuric acid reagent (visible light)

*2.2.4. Antibacterial assay.* The antibacterial assay was performed by CLSI (2012) methods with some modification [15]. The methods was broth microdilution method - Bacteria test used was Staphylococcus aureus with growing medium TSB. A 100 mL of each samples (concentration 15.63-2000 μg/mL), 100 μL TSB and 20 μL suspension of Staphylococcus aureus (incubated 24 hours) were put in well of steril 96-well plates. After that, the mixture in well was incubated at 37 ºC for 24 hours. The minimum concentration of extract at which no visually detectable bacterial growth was described as the minimum inhibitory concentration (MIC). Next, 100 μl of each medium with no visually detectable bacterial growth was inoculated in 100 μl of fresh medium and incubated again for 24 hours at 37 ° C. The minimum concentration at which there was no bacterial growth after the second inoculation was described as the minimum bactericidal concentration (MBC). DMSO solution of 20% used as a negative control and a positive control used tetracycline.

*2.2.5. Antioxidant assay.* The antioxidant activity was determined by two methods, which are microdilution and Thin Layer Chormatography (TLC)-bioautography.
**Microdilution**- 100 μL DPPH solution of 125 μM added to 100 μL of ethanol extract samples with various concentrations of up to a total volume of 200 uL. The mixtures were incubated at 37 °C for 30 minutes. Then, samples absorbance were measured at 517 nm using *Epoch Microplate Spectrophotometer.* Ascorbic acid was used as a positive control. DPPH free radical trapping capacity was calculated [16].

**TLC-bioautography**- Plates that had been developed previously by TLC fingerprint method were sprayed with a solution of 5 mM DPPH and then incubated at 37 ° C for several hours. Yellowish bands appeared indicating that the bands were active as an antioxidant and the plate background was purple. The chromatograms bands were compared with chromatogram that had been detected previously to determine the value of Rf bands that have antioxidant activity [17].

## 3. Results and discussion

Five accessions of *temu hitam* and three other plants (*temu putih*, *temu lawak*, and turmeric) determined in Bogoriense Herbarium, Biology Research Center, LIPI Cibinong. The results showed that five *temu hitam* accessions and three other plants are Zingiberaceae with different species, *Curcuma aeruginosa* Roxb. (Five accession *temu hitam*), *Curcuma zedoaria* (Christm.) Roscoe (*temu putih*), *Curcuma Zanthorriza* Roxb. (*temu lawak*), and *Curcuma longa* L. (turmeric).

The chemical components in medicinal plants relies heavily on the diversity of physiological, harvest condition, plant sources, drying processes, and other factors to ensure the integrity and authenticity of medicinal plants [13,18]. Fingerprint analysis can be used for classification and validation of plant species and multicomponent quality control of medicinal plants. Thin layer chromatography (TLC) was choosen because it provide patterns of fractions of compounds that are specific to each species, beside that the analysis is fast, simple, and relatively cheap [19]. This technique can be applied qualitatively and quantitatively with the presence or absence of a standard compound. Additionally, TLC can provide relatively different patterns of the compound when administered beams with different wavelengths, such as visible light, UV 366 nm and UV 254 nm, as well as the specific dye reagents so as to enrich the detection results (multidetection).

*Temu hitam* samples used in this study came from five different areas, but cultivated in the same specific area so-called as accession. In addition to *temu hitam*, three other plants, turmeric, *temu putih*, and *temu lawak* are used to distinguish component profile between *temu hitam* and the three other plants. By phenotype, *temu hitam* and the three plants have a similar visual appearance, especially among *temu hitam*, *temu putih*, and *temu lawak*, specially on the leaf shape and tinge of purple on the leaf stalk. However, the rhizome can be distinguished when still in the form intact.

Figure 1 shows the profile of compounds in rhizomes, leaves, and stems of five accessions of *temu hitam* and three other plants. All three parts of this plant has a different pattern because it has different type of plant material. The pattern of the compounds in the rhizome is dominated by yellow-green bands and blue in UV 366 nm. Meanwhile, the leaves and stems contain more red bands at UV 366 nm. Different band colors can indicate different compounds. According to Markham (1988), the blue color indicates the presence of flavonoids, flavonon, or flavonols, red as anthocyanidins compound, and greenish as Auron and flavones compound [20]. In UV 254 nm there are only dark bands due to the absorption of the compound to light will diminish and outages, this is because the stationary phase in the form of silica gel containing indicators of fluorescence that can absorb light at a wavelength of 254 nm so that only the stationary phase fluorescent [21]. However, the pattern of the component of three parts plant (rhizomes, stems, and leaves) looks different. In addition to using UV detection 254 and 366 nm, detection is also done with a specific dye reagents. Reagent vanilina-sulfuric acid has the ability to react with chromophore groups on the compound and can enlarge the wavelength (UV-Vis) so that the bands previously not seen in visible light (Figure 1) to be seen in visible light (Figure 2).

The ethanol extract on the same part of the five accessions have similar patterns, but has a different thickness of the bands. Different thickness of the bands can be used as variable comparison because at the time of application all samples have the same concentration and volume and the other conditions was same. The bold band shows abundance or higher concentration of the compound. Blue band with Rf 0.83 is owned by rhizomes five accessions of *temu hitam*, but the intensity of the blue color of the Malang and Cirebon accessions are lower than the three other accessions (Figure 1$a_{iii}$). Then, the band with Rf 0.10 with dark color intensity was lowest for the rhizome Liwa accession, but the intensity of the band Rf 0.55 rhizome is lowest in the Kuningan 1 accession (Figure 1$a_{ii}$). Figure 1 (b) shows the abundance of leaves Liwa accession has the highest compound in the band with Rf 0.70, 0.52, and

IOP Conf. Series: Journal of Physics: Conf. Series **835** (2017) 012014 doi:10.1088/1742-6596/835/1/012014

0.24 in comparison with four other accessions. Meanwhile, the abundance of the highest compound (Rf 0.70 and 0.37) on the stem owned by Cirebon accession (Figure 1c). Based on that, although the five *temu hitam* accessions came from different areas have been cultivated in the same specific area, the abundance or concentration of compounds still have differences.

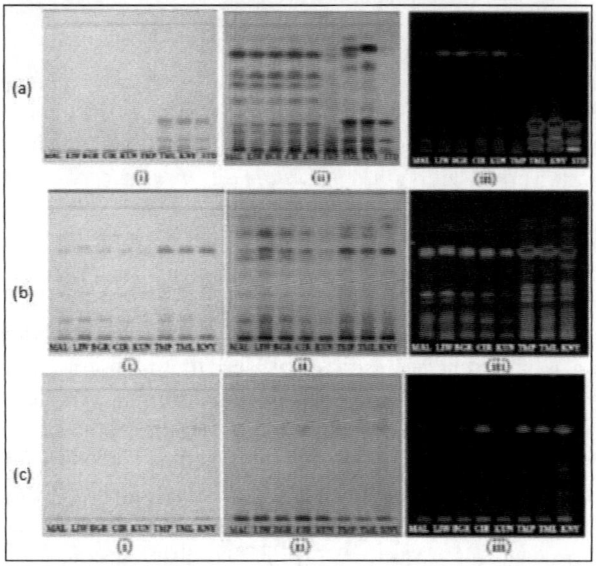


**Figure 1**: The pattern of the fingerprint chromatograms of ethanol extract of rhizome (a), leaves (b), and stem (c) with a mobile phase of chloroform-dichloromethane 9: 1 (v/v). Detection: (i) visible light; (ii) UV 254 nm; and (iii) UV 366 nm. MAL: Malang Accession, LIW: Liwa Accession, BGR: Bogor Accession, CIR: Cirebon Accession, KUN: Kuningan 1 Accession, TMP: Temu Putih, TML: *Temu lawak*, KNY: Turmeric, and STD: Curcuminoid Standard

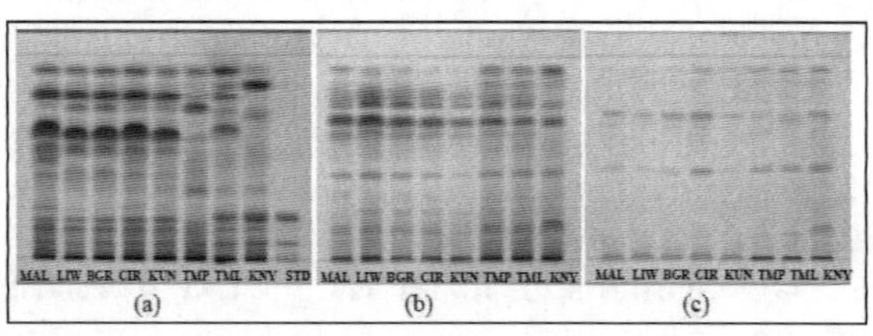

**Figure 2**: The pattern of the fingerprint chromatogram of the ethanol extract of the rhizome (a), leaves (b), and stems (c) with a mobile phase of chloroform-dichloromethane 9:1 (v/v) Detection: vanillin-sulfuric acid visible light. MAL: Malang Accession, LIW: Liwa Accession, BGR: Bogor Accession, CIR: Cirebon Accession, KUN: Kuningan 1 Accession, TMP: Temu Putih, TML: *Temu lawak*, KNY: Turmeric, and STD: Curcuminoid Standard

Ethanol extract profiles of five accessions *temu hitam* difference with the three other plants (Figures 1-2). Rhizomes all plants contain curcuminoid so curcuminoid solution is used as a standard [22-26]. Curcuminoid standard used to view the distribution of the three type of curcuminoid. Three type curcuminoid in Figure 1 and 2a indicated the existence of three bands with Rf 0.22, 0.10, and 0.03. Rhizome five accession *temu hitam* and turmeric contains three bands on a standard but have different thickness intensity, while *temu lawak* contains only band with Rf 0.22 and 0.10 and *temu putih* contain bands with Rf 0.03 and 0.10.

Rhizome of *temu hitam* had characteristic band with Rf 0.42, 0.27, and 0.77 that distinguish it from the rhizome of *temu lawak* and turmeric, but the rhizome of *temu hitam* did not have band with Rf 0.73, 0.20, and 0.64 owned rhizome of *temu lawak* and 0.71 owned turmeric. Band with Rf 0.13 on the rhizome of *temu hitam* is a distinguishing band from rhizome temu putih. In addition, at 254 nm visualization rhizome *temu hitam* did not have band with Rf 0.35 owned turmeric and *temu putih* (Figure 3a). The ethanol extract of leaves of *temu hitam* can be distinguished from leaves three other plants in the band with Rf 0.06 owned *temu hitam* and Rf 0.60 that exist in other plants. Turmeric leaf can be distinguished b y the leaves of *temu hitam*, *temu lawak*, and *temu putih* at 0.23 and 0.96 (contained in turmeric leaves). On the other hand, turmeric leaf did not have band with Rf 0.80 and 0.63 owned *temu hitam*, *temu lawak*, and *temu putih* (Figure 3b). The pattern of chemical components on the stems of *temu hitam* and other plants are not much different from the pattern of the leaves, like Rf 0.94, 0.70, and 0.45 are in both, as well as the characteristic band is same at Rf 0.60 (*temu lawak*, turmeric, and temu putih). However, the number of bands on the leaves more than stems (Figure 3b). Given these distinction bands, *temu hitam* can be discriminated from the three other plants and can avoid counterfeit products are made from *temu hitam* in a commercial application.

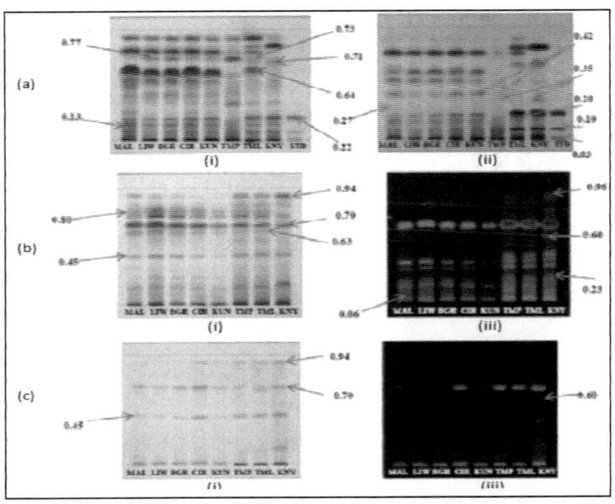


**Figure 3**: Pattern of the fingerprint chromatogram of the ethanol extract of the rhizome (a), leaves (b), and stem (c) with Rf information. Visualization (i) reagent vanillin-sulfuric acid in visible light, and (ii) UV 254 nm with a mobile phase of chloroform-dichloromethane 9: 1 (v/v), (iii) UV 366 nm with a mobile phase of chloroform-dichloromethane 9:1( v/v). MAL: Malang Accession, LIW: Liwa Accession, BGR: Bogor Accession, CIR: Cirebon Accession, KUN: Kuningan 1 Accession, TMP: Temu Putih, TML: *Temu lawak*, KNY: Turmeric, and STD: Curcuminoid Standard

The diversity of components of the profile indicates the potential bioactivity on the plant *temu hitam* the profile shows that *temu hitam* rhizome contains curcuminoid which have a wide range of bioactivity, such as antibacterial and antioxidant [27-29]. *Temu hitam* also contain terpenoids class that reported has antibacterial activity [4,30].

In this study, antibacterial and antioxidant activity of five accessions of *temu hitam* and three other plants were determined bioautography. Based on the results of the ethanol extract of *temu hitam* and other plants have inhibitory activity against *Staphylococcus aureus* with different MIC and MBC. The minimum inhibitory concentration (MIC) called also bacteriostatic, which is the lowest concentration that can inhibit the growth of bacteria. Meanwhile, the concentration that kills 99.9% of bacteria inoculant called minimum bactericidal concentration (MBC). MIC and MBC sample is determined by comparing the turbidity (visual observation) between the sample wells with positive control wells (tetracycline) and negative control (DMSO 20%). Turbid wells indicates there are bacteria grows and components in the sample are able to inhibit the growth of bacteria, while the clear wells indicates components in the sample were able to inhibit even kill bacteria [31].

**Table 2**: The antibacterial activity of ethanol extract of rhizomes, stems, and leaves five *temu hitam* accessions and comparative plants

| Ethanol extracts of *temu hitam*/ comparator samples | MIC (μg/mL) | | | MBC (μg/mL) | | |
|---|---|---|---|---|---|---|
| | R | S | L | R | S | L |
| Kuningan 1 | 250 | 500 | 500 | 500 | * | 2000 |
| Cirebon | 250 | 500 | 1000 | 500 | * | 2000 |
| Malang | 500 | 1000 | 500 | 1000 | * | 2000 |
| Bogor | 500 | 500 | 500 | 500 | * | 2000 |
| Liwa | 250 | 500 | 500 | 1000 | * | 2000 |
| *Temu putih* | 500 | 1000 | 500 | 2000 | * | 2000 |
| Turmeric | 250 | 1000 | 500 | 1000 | * | 1000 |
| *Temu lawak* | 62.5 | 500 | 500 | 250 | * | 2000 |
| Tetracycline | | 15.63 | | | 15.63 | |
| DMSO 20% | | - | | | - | |

(*)= Activity more than 2000 μg/mL; (-)= No activity; R= Rhizomes; S= Stems; L= Leaves

The highest antibacterial activity found in the rhizome followed by leaves and stems (Table 2). Rhizome *temu hitam* Cirebon and Kuningan 1 accession have the highest bacteriostatic and bactericidal activity with MIC and MBC respectively of 250 and 500 μg/mL. However, the antibacterial activity of ethanol extract of rhizome of *temu hitam* Cirebon and Kuningan 1 accession are not higher than the ethanol extract of *temu lawak* rhizome (MIC= 62.5 μg/mL, MBC = 250 μg/mL) and tetracycline as a positive control (MIC=MBC= 15.63 μg/mL).

**Table 3.** $IC_{50}$ value of *temu hitam* ethanol extract, comparative plants, and ascorbic acid

| Accessions of *temu hitam*/ comparator samples | $IC_{50}$ (μg/mL) | | |
|---|---|---|---|
| | Rhizomes | Stems | Leaves |
| Cirebon | 131.40 | 315.81 | 193.84 |
| Malang | 131.56 | 561.25 | 163.05 |
| Bogor | 130.01 | 702.00 | 194.63 |
| Liwa | 124.88 | 573.38 | 160.58 |
| Kuningan 1 | 146.71 | 481.46 | 211.49 |
| *Temu putih* | 94.35 | 1335.20 | 162.60 |
| Turmeric | 18.82 | 417.93 | 171.36 |
| *Temu lawak* | 18.45 | 305.06 | 157.17 |
| Ascorbic acid | | 3.22 | |

Various compounds contained in the five accessions *temu hitam* profile makes this plant has a variety of activities, not only antibacterial activity, but other activities, such as antioxidants. The ethanol extract five accession *temu hitam* on different parts have different relative $IC_{50}$ values, the concentration that inhibit 50% radicals (Table 3). Based on the $IC_{50}$ value, the rhizome five accessions *temu hitam* had the highest activity than stems and leaves, but its activity is lower than turmeric, *temu lawak* and temu putih. This is also supported by the profile component of the rhizome that has a number of compounds (band) most. $IC_{50}$ value rhizomes and leaves of Liwa accession has the highest antioxidant activity than the other accessions, but on the stem, Cirebon accession which has the highest activity.

IOP Conf. Series: Journal of Physics: Conf. Series **835** (2017) 012014 doi:10.1088/1742-6596/835/1/012014

Besides determined by $IC_{50}$ value, the antioxidant activity was also determined qualitatively using TLC-bioautography. In this method showed yellowish specific band with a purple background plate, yellow band emerged due to the reduction of DPPH and indicates that the band is active of antioxidants. Yellow band thickening showed stronger antioxidant activity [17]. Results showed *temu hitam* and comparator plants rhizomes have thickest yellow bands and most visible than the leaves and stems. Rhizomes of Liwa and Bogor accessions have higher activity, this is due to the accessions of these have yellow bands with Rf 0.77 and 0.42 thicker than the other accessions (Figure 4a). Likewise, the leaves and stems, leaf of Liwa accession and stem of Cirebon accession showed yellow bands higher intensity than the other accessions (Figure 3a). These results are also supported by the results of fingerprint chromatogram pattern in Figure 1-2.

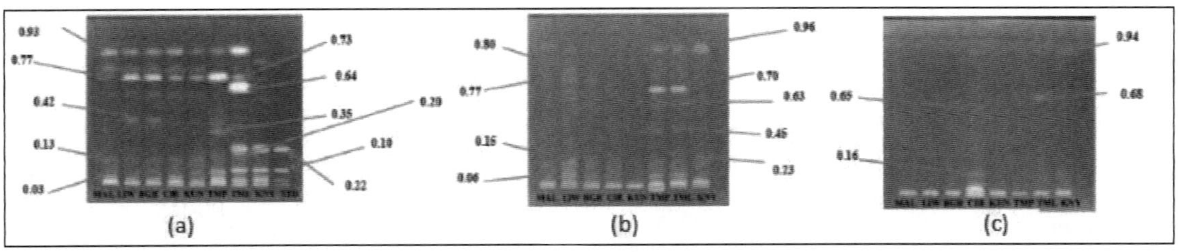


**Figure 4.** TLC-bioautogram of ethanol extract of rhizomes (a), leaves (b), and stems (c) with chloroform:dichloromethane 9:1 (v/v) as mobile phase. Detection by DPPH 5mM in visible light

At the five *temu hitam* accessions, bands with active antioxidant are Rf of 0.93, 0.83, 0.77, 0.52, 0.42, 0.10, 0.13. and 0.03, while Rf of 0.77, 0.83, and 0.93 are most active. Meanwhile, the most active rhizome bands of *temu lawak* are 0.64, 0.22, 0.10, and 0.93. Turmeric has active band similar to *temu lawak*, just do not have active band at Rf 0.64 and 0.93, but there are active bands at Rf 0.88, 0.10. The thickest band at 0.77 found in *temu putih* and have more active band at 0.35 and 0.42 (Figure 4b). On leaves and stems have less active bands of antioxidants than the rhizome. This is also evidenced by the $IC_{50}$ value of the leaves and stems lower than the rhizome.

The leaves of *temu hitam* has active band at Rf of 0.80, 0.94, 0.77, 0.70, 0.15 and 0.06. However, active bands are more clearly visible only in Liwa accession. Furthermore, characteristic band at 0.96 of turmeric has activity, whereas in *temu lawak* and *temu putih* the most active bands are 0.70 and 0.45 (Figure 4b). Stems have very weak antioxidant activity seen from the active band only seen in *temu hitam* Cirebon accession, *temu lawak*, and turmeric at 0.94 and 0.16, 0.65 (*temu hitam* Cirebon), and 0.68 (*temu lawak*) (Figure 4c).

The antioxidant activity of ethanol extract of *temu hitam* rhizome lower than *temu lawak*, *temu putih*, and turmeric rhizome. It can be caused by the content of existing curcuminoid. Curcuminoid can be used as one factor for the level of antioxidant because three type of curcuminoid is active as an antioxidant. Based on Figure 4a curcuminoid content of the ethanol extract of *temu hitam* rhizome lower than the other three rhizomes, especially in turmeric and *temu lawak* rhizome. These results are also consistent with research Jitoe *et al.* (1992) which states curcuminoid content of *temu lawak* and turmeric rhizome higher than *temu hitam* [32]. In the bioautogram shown a yellow bands curcuminoid at Rf of 0.10 and 0.03 on *temu hitam* thinner even at 0.22 almost no activity (low intensity). Additionally, *temu lawak* has an additional activity from Rf of 0.20, 0.64 and 0.73 bands and turmeric from Rf of 0.88. *Temu putih* has bands at Rf of 0.77 and 0.10 thicker than *temu hitam*. Bioautogram shown antioxidant activity is directly proportional to the $IC_{50}$ value generated by the accession Liwa rhizomes that have the highest activity, but lower than *temu lawak*, turmeric, and *temu putih*.

## 4. Conclusion

The ethanol extracts of five accessions *temu hitam* have different chromatographic fingerprints profiles between rhizomes, stems and leaves, but has a similar profile to the same part of plant.

Rhizome of five accession *temu hitam* and turmeric contains three bands of standard of curcuminoid with the different thickness intensity, while *temu lawak* contains only band with Rf 0.22 and 0.10 and *temu putih* with Rf 0.03 and 0.10. Rhizomes of *temu hitam* has characteristic bands at Rf 0.42, 0.27, and 0.77 that distinguish it from the rhizome of *temu lawak* and turmeric, and Rf 0.13 with *temu putih* rhizomes. Leaves and stems of five accessions *temu hitam* and comparator plants has similar fingerprint profile and band at 0.60 distinguishing between the stem and leaves of *temu hitam* with the leaves and stems of comparator plants.

Rhizomes of *temu hitam* and comparator plants have the highest antibacterial and antioxidant activity than their leaves and stems. The antibacterial activity of Cirebon and Kuningan 1 accessions rhizomes are stronger than the other accessions but lower than the rhizome of *temu lawak* and tetracycline (positive control). Based on the $IC_{50}$ and TLC-bioautography, ethanol extract of rhizomes Liwa accession has strongest antioxidant activity than the other accessions, but lower than *temu lawak*, turmeric, and *temu putih*.

**References**

[1] Philip K, Malek S N A, Sani W, Shin S K, Kumar S, Lai H K, Serm L G and Rahman S N S A 2009 *Am. J. Appl. Sci.* **6** 1613-7
[2] Reanmongkol W, Subhadhirasakul S, Khaisombat N, Feungnawakit P, Jantasila S and Khamjun A 2006 *Songklanakarin J. Sci. Technol.* **28** 999-1008
[3] Hossain C F, Al-Amin M, Sayem A S Md, Siragee I H, Tunan A M, Hassan F, Kabir Md M, Sultana G N H 2015 *BMC Complement. Altern. Med.* **15** 1-7
[4] Kamazeri Tg S A Tg, Samah O A, Taher M, Susanti D and Qaralleh H 2012 *Asian Pac. J. Trop. Dis.* 202-9
[5] Tambunan D Y S 2014 *Minyak atsiri rimpang, batang, dan daun temu hitam (Curcuma aeruginosa Roxb.) sebagai antibakteri Streptococcus mutans dan pendegradasi biofilm pada gigi* (Bogor: Bogor Agricultural Univesity)
[6] Angel G R, Vimala B and Nambisan B 2012 *IJPPR* **4** 69-73
[7] Nurcholis W, Khumaida N, Muhamad S, Maria B and Ardyani 2015 *Int. J. Res. Ayurveda Pharm.* **6** 634-7
[8] Theanphong O, Mingvanish W and Kirdmanee C 2015 *Bull. Health Sci. Tech.* **13** 6-16
[9] Mazid M, Khan T A and Mohammad F 2011 *Biol. Med.* **3** 232-49
[10] Scafer H and Wink M 2009 *Biotechnol. J.* **4** 1684-703
[11] Wink M 1988 *Theor Appl Genet.* **75** 225-33
[12] Septaningsih D A 2015 *Identifikasi Komponen Temu Ireng (Curcuma aeruginosa) dengan Pola Sidik Jari Kromatografi Cair-Spektroskopi Massa dan Kemometrik* (Bogor: Bogor Agricultural Univesity)
[13] Figueiredo A C, Barroso J G, Pedro L G and Scheffer J J C 2008 *Flavour Frag. J.* **23** 213-26
[14] Borges C N, Bruns R E, Almeida A A and Scarminio I S 2007 *Anal. Chim. Acta* **595** 28-37
[15] Cockerill F R, Wikler M A, Alder J, Dudley M N, Eliopoulos G M, Ferraro M J, Hardy D J, Hecht D W, Hindler J A, Patel J B, Powell M, Swenson J M, Thomson R B, Traczewski M M, Turnidge J D, Weinstein M P and Zimmer B L 2012 *Methods for Dilution Antimicrobial Susceptibility Tests for Bacteria That Grow Aerobically; Approved Standard Ninth Edition* vol 32 (Wayne, Pennsylvania: Clinical and Laboratory Standards Institute)
[16] Batubara I, Mitsunaga T, and Ohashi H 2009 *J. Wood Sci.* **55** 230-5
[17] Salazar-Aranda R, Perez-Lopez L A, Loppez Arroyo J, Alanis-Garza B A, and Torres N W 2009 *Evid. Based Complement. Alternat. Med.* **2011** 1-6
[18] Liang Y Z, Xie P, and Chan K 2004 *J. Chromatogr. B* **812** 53-70
[19] Liu W J H 2011 *Traditional Herbal Medicine Research Methods: Identification, Analysis, Bioassay, and Pharmaceutical and Clinical Studies* (New Jersey: John Wiley & Sons, Inc.)
[20] Markham K R. 1988 *Techniques of Flavonoid Identification*, (New York : Academic Press)

[21] Sherma J and Fried B 2003 *Handbook of Thin-Layer Chromatography* (New York: Marcell Dekker Inc.)
[22] Lobo R, Prabhu K S, Shirwaikar A, Shirwaikar A 2009 *J. Pharm. Pharmacol.* **61** 13-21
[23] Mau J, Lai E Y C, Wang N, Chen C, Chang C, Chyau C 2003 *Food Chem.* **82** 583-91
[24] Jayaprakasha G K, Rao L J M, Sakariah K K 2002 *J. Agric. Food Chem.* **50** 3668-72
[25] Lechtenberg M, Quandt B, Nahrstedt A 2004 *Phytochem. Anal.* **15** 152-8
[26] Bos R, Windono T, Woerdenbag H J, Boersma Y L, Koulman A and Kayser O 2007 *Phytochem. Anal.* **18** 118-22
[27] Kowsalya R and Krishnaveni M 2011 Ex *J. Pure Appl. Microbiol.* **5** 317-21
[28] Sivasothy Y, Sulaiman S F, Ooi K L, Ibarhim H and Awang K 2013 *Food Control* **30** 714-21
[29] Jayaprakasha G K, Rao L J and Sakariah K K 2006 *Food Chem.* **98** 720-4
[30] Simoh S and Zainal A 2015 *Asian Pac. J. Trop. Biomed.* **5** 412-7
[31] Basri D F and Fan S H 2005 *Indian J. Pharmacol.* **37** 26-9
[32] Jitoe A, Masuda T, Tengah I G P, Suprapta D N, Gara I W and Nakatani N 1992 *J. Agric. Food Chem.* **40** 1337-40

# Study of DNA adduct 8 hydroxy-2'-deoxyguanosine (8-OHdG) formation through fenton reaction with tert-butylhydroquinone (TBHQ) and butyl hydroxy toluene (BHT)

**S Handayani, I C Dani, Budiawan* and D Pakuanisa**

Department of Chemistry, Faculty of Mathematics and Natural Sciences Universitas Indonesia, Depok 16424, Indonesia

E-mail : dr.budiawan@gmail.com

**Abstract**. The research of DNA adduct formation 8-hydroxy-2'-Deoxyguanosine (8- OHdG) as a biomarker of DNA damage due to oxidative stress was carried out by reacting the DNA base 2'-deoxyguanosine-5'-monophosphate with TBHQ and BHT. The formationof 8-OHdG was carried out in various conditions,at temperature of 37° C and 60° C, pH 7.4 and pH 8.4, within 5 hours of incubation time and in the addition of $FeSO_4$.The formation of DNA adducts profile were analyzed using reversed phase HPLC with UV detector at a wavelength of 254 nm. The results of the study showed that TBHQ and BHT can trigger the formation of 8-OHdG from the reaction of 2'-hydroxy Deoxyguanosine-5'-monophosphate in the presence of Fe (II). Meanwhile, in the addition of hydrogen peroxide, the formation of DNA adducts only occur in the test substance TBHQ. The results showed that the condition of higher temperature at 60°C and pH 8,4 affects the higher formation of DNA adducts.

## 1. Introduction

Food is a basic needed which is essential for human life and the most important or necessary things in order to survive. Generally in food processing, treatment are given by various ways such as adding additional materials in order to extend the shelf life, improving the texture, or appearance of delicacy[1]. A safe food is a food that does not contain (1) biological hazard or microbiology hazard, (2) chemical hazards, and (3) physical hazard [2].

Antioxidants are natural antioxidants that have been produced synthetically for commercial purposes [3]. Synthetic antioxidants are commonly used for food additives to prevent rancidity caused by fat oil oxidation in food. Butyl hydroxyanisole (BHA), butyl hydroxy toluene (BHT), propyl gallate (PG), Tert-Butyl Hidoksi Quinon (TBHQ) and tocopherols are allowed to be used in foods. Antioxidant BHT in the form of white crystalline solid is widely used because its relatively inexpensive [3]. Antioxidant TBHQ is known as the most effective antioxidant for fats and oils, particularly oil plant because it has antioxidant abilities both in the frying process, but low in the combustion process. Antioxidant TBHQ known shaped white to light brown powder, has a sufficient solubility in fats and oils, not forming complexes with Fe and Cu colors but can turn to be pink in the presence of a base [3]. Acceptable daily intake for TBHQ is 0.70 mg/kg body weight [4-5].

* Corresponding author

In abnormal biological conditions, there is always the potential for an antioxidant to become a pro-oxidant if a suitable receptor molecule presents to accept the electron and promote the autoxidation. Mineral ions are particularly important as pro-oxidants [6]. Okubo, et.al 1997, reported that TBHQ can cause the breakdown of DNA in vitro and form 8-hydroksi-2'-deoksiguanosine as a result of oxidative DNA damage in Calf Thymus DNA [7]. BHT-quinone (2,6-di-tert-butil-p-benzoquinone), has been reported to cause cleavage of supercoiled DNA [8]. Oikawa et al. reported that BHT metabolites, i.e., BHT-quinone and 2,6-di-tert-butyl-4-hydroperoxyl-4-methyl-2,5-cyclohexa-dienone (BHT-OOH), induced oxidative DNA damage [9].

Furthermore, this study was conducted to determine the formation of 8-hydroxy-2'-deoksiguanosin (8-OHdG), which is an indicator of DNA damage due to chemical carcinogens exposure. In this study 2'-deoxyguanosine-5'-monophosphate is used to be reacted with BHT and TBHQ in the various pH and temperature and the addition of Fe (II) to see the profile of 8-OHdG formation and prove that carcinogen compounds exposure can cause DNA damage.

## 2. Experimental and Method

In this research, the study conducted were in vitro study of DNA base 2'-deoxyguanosine and antioxidants trigger to DNA Adduct formation by using HPLC, the analytical method has been validated in the previous research.

Study of DNA binding interaction was observed by incubating 350 μL of 2'-deoxyguanosine 5'-monophosphate (500 ppb) in potassium phosphate buffer 0,1 M (pH 7.4 and 8.4) with 350 μL of TBHQ (5000 ppb) and 350 μL of aquabide st for 5 hours at variation of temperature 37°C and 60°C. The variation condition undergo in the addition of 350 μL of $FeSO_4$ (5000 ppb), using BHT (5000 ppb) and under Fenton reaction using 175 μL of $FeSO_4$ (10000 ppb), and 175 μL of $H_2O_2$ (10000 ppb). All variation were shown in Table 1. The incubation products were analyzed by using HPLC with analysis condition shown in Table 2.

**Table 1** Variation mixture and condition of dGMP incubation

| Variation mixtures | | | |
|---|---|---|---|
| In the presence of $H_2O_2$ | | without $H_2O_2$ | |
| Temperature incubation 37°C | Temperature incubation 60°C | Temperature incubation 37°C | Temperature incubation 60°C |
| dGMP pH 7,4 + TBHQ + $H_2O_2$ | | dGMP pH 7,4 + TBHQ | |
| dGMP pH 8,4 + TBHQ + $H_2O_2$ | | dGMP pH 8,4 + TBHQ | |
| dGMP pH 7,4 + BHT + H2O2 | | dGMP pH 7,4 +BHT | |
| dGMP pH 8,4 + BHT + H2O2 | | dGMP pH 8,4 + BHT | |
| dGMP pH 7,4 + FeSO4 + H2O2 | | dGMP pH 7,4 + FeSO4 | |
| dGMP pH 8,4 + FeSO4 + H2O2 | | dGMP pH 8,4 + FeSO4 | |
| dGMP pH 7,4 + TBHQ + FeSO4 + H2O2 | | dGMP pH 7,4 + TBHQ + FeSO4 | |
| dGMP pH 8,4 + TBHQ+ FeSO4 + H2O2 | | dGMP pH 8,4 + TBHQ+ FeSO4 | |
| dGMP pH 7,4 + BHT+ FeSO4 + H2O2 | | dGMP pH 7,4 + BHT+ FeSO4 | |
| dGMP pH 8,4 + BHT + FeSO4 + H2O2 | | dGMP pH 8,4 + BHT + FeSO4 | |

**Table 2** Analysis condition using HPLC

| | |
|---|---|
| Colomn | X Bridge C 18 Waters (4,6 X 250 mm, particle size 5 μm) |
| Colomn temperature | 31°C |
| Eluent | Sodium hydrogen phosphate buffer 10 mM and Methanol (85 : 15) |
| Flow rate | 1,0 mL/minute |
| Volume Injection | 20μL |
| Detector | UV wavelength 254 nm |

## 3. Result and Disscussion

3.1 Result of Adducts

This study was conducted by incubated the nucleoside of deoxyguanosine with TBHQ and BHT. This reaction was assumed to produce adducts 8-OHdG. The 8-OHdG compounds are formed from a radical hydroxyl radical (•OH) attack on basic groups of guanine in the C8 atom [10]. The changing of deoxyguanosine structure becomes 8-OHdG due to oxidative stress can be seen in Figure 1.

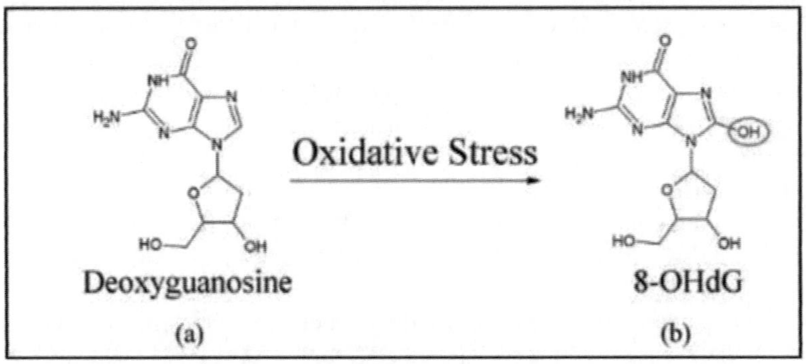


**Figure 1** (a) Deoxyguanosine structure ; (b) 8-OHdG structure, deoxyguanosine modification [10]

3.2 The Effect of Xenobiotics

The reaction between DGMP with TBHQ and BHT which are xenobiotic trigger to radical formation,was undergo, to see the result of 8-OHdG formation which indicate the presence of DNA damage when the human body has been exposed to xenobiotics contained in the antioxidant artificial in food such as TBHQ and BHT.

*3.2.1. The Effect of Compounds Trigger to Free Radicals $Fe^{2+}$ , and $H_2O_2$.* The reaction of the DGMP with BHT or TBHQ in the presence of $Fe^{2+}$ have different results. At the temperature of 37°C and at pH 7.4 for TBHQ and BHT reaction, the formation of adducts were not detected. However, in another variation on the test xenobiotics substance TBHQ and BHT at pH 8.4 at temperature of 37°C (Figure 2), the adducts were detected as well as the variation reaction at the temperature of 60°C with pH 7.4 and 8.4.

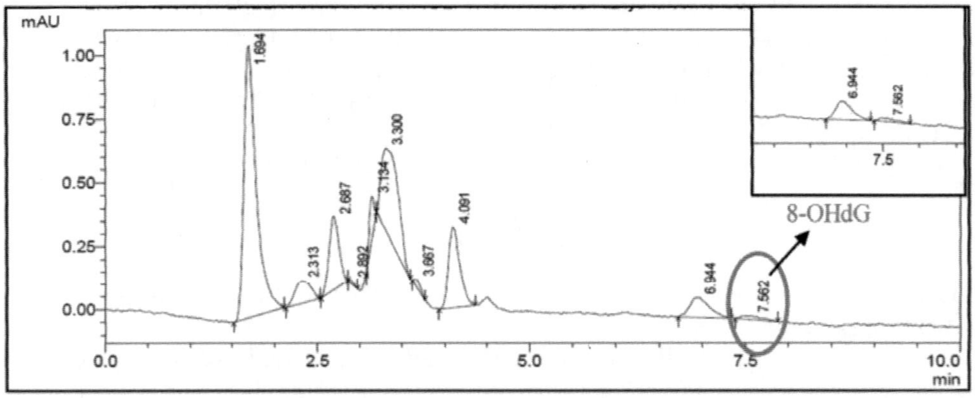


**Figure 2.** Chromatogram of dGMP and BHT in the presence of $Fe^{2+}$ in the temperature 60°C and pH 8,4

The addition of $Fe^{2+}$ is according to its contained in the body particularly containing as the active site of CYP450 enzymes. The $Fe^{2+}$ can trigger to the formation of radicals in its oxidized form, Fe (III), through Fenton reaction:

$$H_2O_2 + Fe^{2+} \longrightarrow Fe^{3+} + \bullet OH + OH^-$$

The differences of 8-OHdG levels shown in the variation of the test substances. The formation of adducts was not detected from TBHQ and BHT reaction in the addition of $Fe^{2+}$ at pH 7.4 and at temperature 37°C. Meanwhile, at pH 8.4, the adducts were detected as much as 6.90 ppb for TBHQ reaction and 6.90 ppb for BHT reaction. The graphic of DNA adduct formation at temperature 37°C as the effect of $H_2O_2$ presence has shown in Figure 3.3. At the temperature of 60°C with the same conditions, the level of 8-OHdG were measured as much as 6.74 ppb at pH 7.4 and 7.86 ppb at pH 8.4 with the test substance TBHQ and in the addition of $Fe^{2+}$. On the test substance of BHT, the DNA Adduct 8-OHdG at pH 8.4 was measured as much as 10.22 ppb, but at pH 7.4 the DNA adduct formation was not detected (Figure 3).

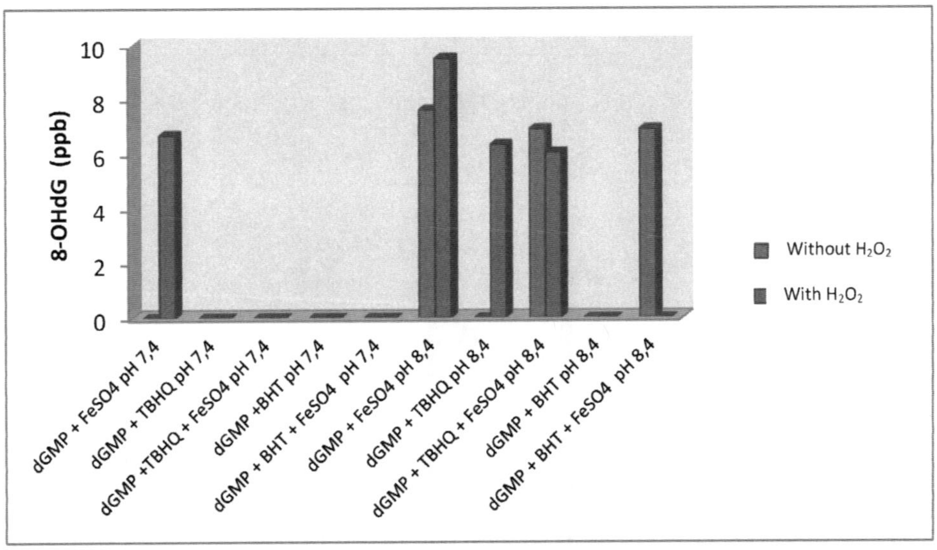


**Figure 3.** The effect of $Fe^{2+}$ and $H_2O_2$ in DNA Adduct 8-OHdG formation in the temperature of 37°C

The result of the reaction by the addition of $H_2O_2$ to the test substance TBHQ without the presence of $Fe^{2+}$ showed the formation of DNA Adducts 8-OHdG adduct which only detected at temperature of 37°C and pH 8.4 as much as 6.32 ppb. Meanwhile, at temperature of 60°C the DNA Adduct 8-OHDG was detected as much as 7.09 ppb at pH 7.4 and 8.21 ppb at pH 8.4. It was different from BHT reaction, which has produced the DNA adduct formation only at temperature of 60°C at both pH as much as 7.38 ppb at pH 7.4 and 7.34 ppb at pH 8.4.The graphic of DNA Adduct level in the temperature of 60°C in the presence and without a presence of $H_2O_2$ has shown in Figure 4.

*3.2.2 The Effect of Fenton Reaction.* The addition of $H_2O_2$ is an important factor in the reaction of this study. The results of the test substance incubation of DGMP and TBHQ or BHT in the addition of $Fe^{2+}$ and in the variation of pH 7.4 and 8.4 and temperature of 37°C and 60°C, showed the variation results of adducts. The results of analysis of 8-OHdG levels in the reactions using the test substance TBHQ and $Fe^{2+}$ at pH 8.4 and temperature 37°C was 6.03 ppb, meanwhile at pH of 7.4 the 8-OHdG was not detected. At the same condition of incubation but in the temperature variations of 60°C, showed different results. The 8-OHdG levels at pH 7.4 was 6.01 ppb and at pH 8.4 was 6.81 ppb. The graphic

IOP Conf. Series: Journal of Physics: Conf. Series **835** (2017) 012015 doi:10.1088/1742-6596/835/1/012015

of DNA adduct formation from the effect of Fenton reaction in the presence of TBHQ has shown in Figure 5.

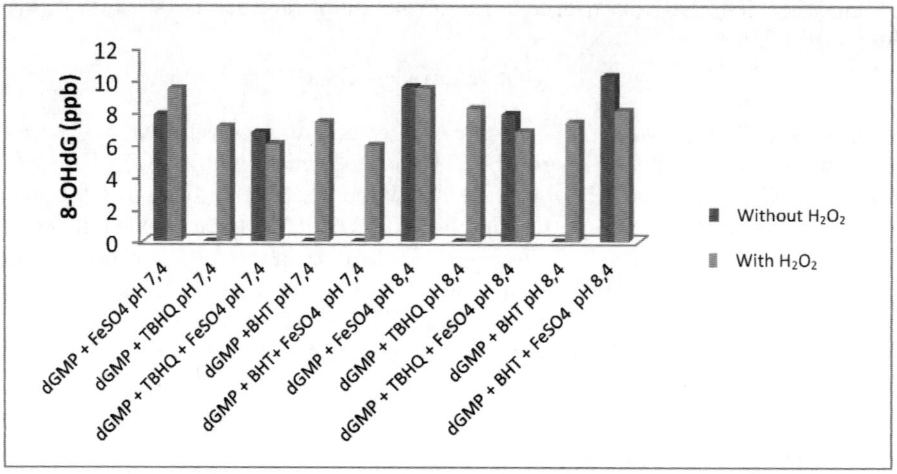


**Figure 4.** The effect of $Fe^{2+}$ and $H_2O_2$ towards the formation of DNA Adducts 8-OHdG in the temperature of 60°C

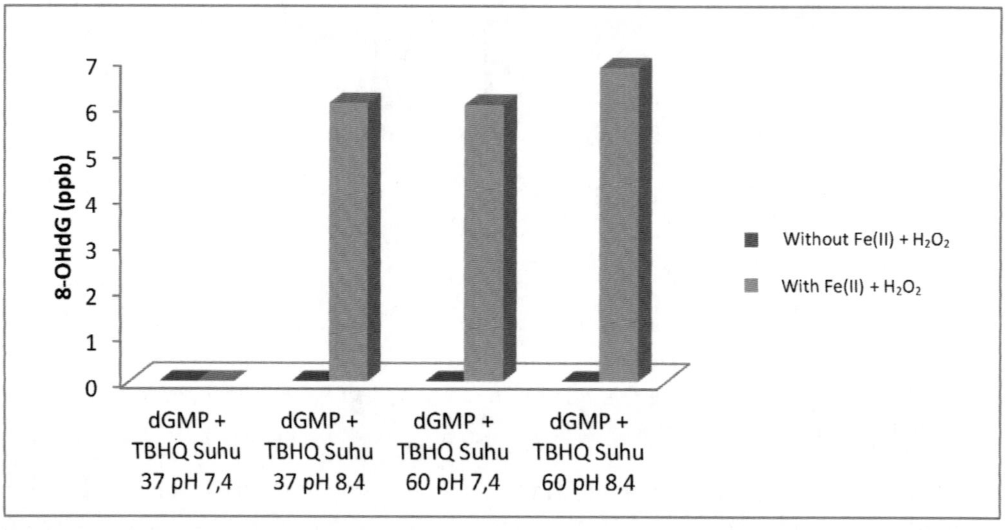


**Figure 5**. The effect of Fenton reaction towards DNA Adduct 8-OHdG Formation with TBHQ

On the incubation using BHT and $Fe^{2+}$ at temperature 37°C, 8-OHdG level at pH 7.4 and 8.4 was not detected. At a temperature of 60°C both of variation at pH 7.4 and pH 8.4 was detected the level of 8-OHdG as much as 5.94 ppb for pH 7,4 and 8.09 ppb for pH 8.4. The graphic of DNA Adduct formation in the presence of BHT has shown in Figure 6.

The differences of 8-OHdG levels is due to the influence of $H_2O_2$ which has a function as an oxidator. Oxidative DNA damage caused the formation of hydroxyl radicals from $H_2O_2$ through the Fenton reaction (Tokoyuni and Sagripanti, 1996). The $H_2O_2$ presences in the reaction process will form the oxidation reaction of $Fe^{2+}$ to $Fe^{3+}$ and produce hydroxyl radical (•OH) which have a reactivity and will interact with cells to form adducts [12]. The hydroxyl (OH•) radical is a reactive ROS which will react with DNA and forming an adduct 8-OHdG as a marker of oxidative DNA damage [13]. The reaction that occurs without the addition of $H_2O_2$ will lead to autoxidation process. Autoxidation is a

reaction that is triggered by the air on organic and inorganic systems which are actually the result of directly or indirectly reaction from molecular oxygen in fre radical process (14).
The formation of adducts in the presence of fenton reaction have been influenced and have synergistic effect either with TBHQ or BHT especially at pH 8.4 and temperature 60°C.

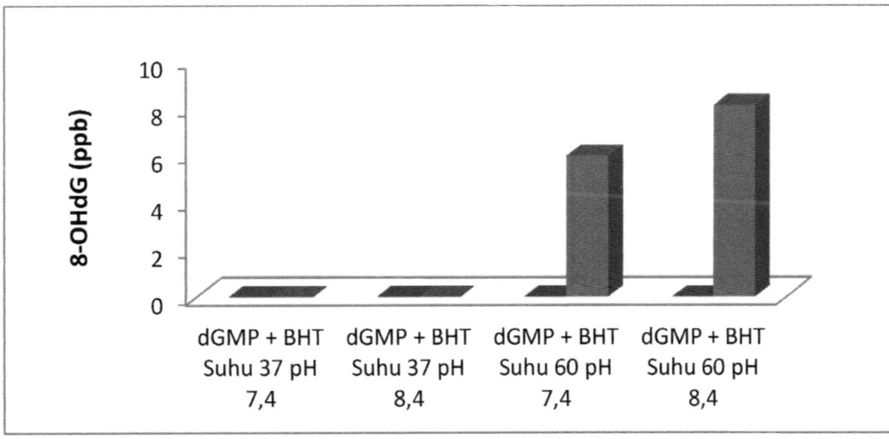


**Figure 6.** The effect of Fenton Reaction towards 8-OHdG formation with BHT

## 4. Conclusion

The reaction of Deoxyguanosine monophosphate either with TBHQ or BHT in the presence of $Fe^{2+}$ at temperature 60°C and at pH 8.4 was affected the formation of 8-OhdG. At the temperature of 37°C the formation of adducts eiher with TBHQ, or BHT in the presence of $Fe^{2+}$ was detected only at pH 8.4. The result of the reaction by the addition of $H_2O_2$ to the test substance TBHQ without $Fe^{2+}$ produces adduct which only detected at temperature of 37°C and pH 8.4 meanwhile at temperature of 60°C and at both pH 7.4 and 8.4 8-OHdG was formed. The result of the reaction by the addition of $H_2O_2$ to the test substance BHT without $Fe^{2+}$ produces adduct at temperature of 60°C and at both pH 7.4 and 8.4. The Fenton reaction was influenced the formation of DNA Adduct 8-OHdG by the condition of temperature and pH. At temperature 37°C, the 8-OHdG was detected only in the presence of TBHQ at pH 8.4. At temperature 60°C, the 8-OHdG was detected in the presence of TBHQ or BHT at both pH 7.4 and 8.4. Finally, the formation of adduct at temperature 60°C was higher than at temperature 37°C

## References

[1] De Vries, J 1997 *Food and Toxicity* (New York : CRC Press)
[2] Ardiansyah 2006 Keamanan Pangan Fungsional Berbasis Pangan Tradisional retrieved from http:/ /www.beritaiptek.com
[3] Buck 1991 *Antioxidant in Food Additive* User's Hand Book. ed JimSmith (eds) (London: Blackie & Sons Ltd) pp 149-83
[4] JECFA 1994 (2016, January 4) retrieved from http://www.fao.org/ag/agn/jecfa-additives/specs/Monograph1/Additive-459.pdf
[5] European Food Safety Authority (EFSA) 2012 *EFSA Journal* **10** 258.
[6] FAO/WHO 2004 *Vitamin and Mineral Requirements in Human Nutrition.* Second Edition. Chapter 7, (Bangkok: WHO Library Cataloguing-in-Publication Data) p130–9.
[7] Okubo T, Nagai F, Ushiyama K and Kano I 1997 *Toxicol. Lett.* **90** 11–8
[8] Nagai F, Ushiyama K and Kano I 1993 *Arch. Toxicol.* **67** 552-7
[9] Oikawa S, Nishino K, Oikawa S, Inoue S, Mizutani T and Kawanishi S 1998 *Biochem. Pharmacol.* **56** 361-70
[10] Commodore A, Zhang J Chang Y, Hartinger S M, Lanata C F, Mäusezahl D, Gil A I, Hall D

B, Aguilar-Villalobos M, Vena J E, Wang J S, and Naeher L P 2013 *Environ Int.* **60** 112-22
[11] Tokoyuni S and Sagripanti J L 1996 *Free Radic. Biol. Med.* **20** 859-64
[12] Rachmawati and Widya 2014 *Studi Pembentukan 8-hidroksi 2'-deoksiguanosin (8-OHdG) dari 2'deoksiguanosindan Guanosin dengan Paparan Senyawa Tert-Butil Hidrokuinon dan Katalis* $TiO_2$ *Sebagai Biomarker Risiko Kanker* (Depok:Universitas Indonesia)
[13] Valavanidis A, Vlachogianni T and Fiotakis C 2009 *J Environ Sci Health* C **27** 120–39
[14] Simic M G 1981 *J.Chem.Educ* **58** 125-31

IOP Conf. Series: Journal of Physics: Conf. Series **835** (2017) 012016 doi:10.1088/1742-6596/835/1/012016

# High perfomance liquid chromatography fingerprint analysis for quality control of brotowali (*Tinospora crispa*)

**V B Syarifah[1], M Rafi[1,2*], W T Wahyuni[1,2]**

[1] Department of Chemistry, Faculty of Mathematics and Natural Sciences, Bogor Agricultural University, Kampus IPB Darmaga, Bogor, 16680, Indonesia
[2] Tropical Biopharmaca Research Centre – Research and Community Empowerment Institute, Bogor Agricultural University, Kampus IPB Taman Kencana Bogor 16128, Indonesia

E-mail : mra@ipb.ac.id

**Abstract**. Brotowali (*Tinospora crispa*) is widely used in Indonesia as ingredient of herbal medicine formulation. To ensure the quality, safety, and efficacy of herbal medicine products, its chemical constituents should be continuously evaluated. High performance liquid chromatography (HPLC) fingerprint is one of powerful technique for this quality control process. In this study, HPLC fingerprint analysis method was developed for quality control of brotowali. HPLC analysis was performed in C18 column and detection was performed using photodiode array detector. The optimum mobile phase for brotowali fingerprint was acetonitrile (ACN) and 0.1% formic acid in gradient elution mode at a flow rate of 1 mL/min. The number of peaks detected in HPLC fingerprint of brotowali was 32 peaks and 23 peaks for stems and leaves, respectively. Berberine as marker compound was detected at retention time of 20.525 minutes. Evaluation of analytical performance including precision, reproducibility, and stability prove that this HPLC fingerprint analysis was reliable and could be applied for quality control of brotowali.

## 1. Introduction

Brotowali (*Tinospora crispa*) is a medicinal plant that commonly used as ingredient of antidiabetic formula. This medicinal plant contains several types of secondary metabolites, including quaternary alkaloids [1], diterpenoid lactones, flavanoid glycosides, steroid, sesquiterpenoids, phenolic, and polysaccharide [2]. Those secondary metabolites are expected to be related with pharmacological activities of brotowali as antihyperglycemic [3], antioxidant, antiproliferative [4], antimalarial [5], antimicrobe [6], and anti-inflammatory [7].

The content of chemical constituents, including secondary metabolites, in brotowali could be affected by some factors, such as cultivating area, climate (temperature, humidity, light, and wind), geography, harvest time, processing methods, and storage. Since the secondary metabolites of brotowali (individual or synergistic) expected to be correlated with its pharmacological activities, evaluation of chemical constituents of brotowali could be performed as quality control of brotowali as raw material of herbal medicine.

* Corresponding author

The pattern-oriented approach such as fingerprinting is a useful technique for quality control of herbal medicine and its raw materials. A comprehensive profile of all chemical constituents of a herbal medicine could be exhibited in fingerprints profile, make it convenient for identification and authentication of raw materials and herbal products [8,9]. Chromatography fingerprint analysis has been developed as a method of quality control and also recommended by the World Health Organization (WHO), the Food and Drug Administration (FDA), European Medicines Agency (EMA), and the Chinese State Food and Drug Administration (CSFDA) [10]. High performance liquid chromatography (HPLC) is an analytical technique that is widely used for quality control method based on chromatography. HPLC has several advantages in terms of analysis, which is more efficient, high sensitivity, high accuracy, good repeatability, and can identify chemical compounds in a material in all measurements [11].

In this study, HPLC fingerprint analysis of brotowali was developed using reverse phase chromatography system. Berberine was used as marker compound since this compound is one of the important quaternary alkaloids in brotowali. Berberine was reported as inhibitors of acetylcholinesterase [12] and as antidiabetic [13].

## 2. Research Method

### 2.1. Material

Brotowali (leaves and stems), binahong (*Anredera cordifolia*) leaves, green betel (*Piper betle*) leaves were obtained from Tropical Biopharmaca Research Center garden located at Cikabayan, Bogor, Indonesia. Berberine with the purity of ≥90% (Sigma-Aldrich, Missouri, United States), ethanol, HPLC grade methanol, acetonitrile HPLC grade, formic acid (Merck, Darmstadt, Germany), and aquadest.

### 2.2. Sample and Standard Preparation

500 mg of dried and powdered sample were sonicated with 5 mL of methanol for 1 hour at room temperature. After filtration through a 0.22 μm membrane filter, the sample solution diluted to 10 mL with methanol before injected into HPLC system. Berberine solution was prepared in methanol at concentration of 100 ppm. In other flask, sample solution was spiked with 1 mL of berberine standard (100 ppm) and diluted to 10 mL with methanol.

### 2.3. Chromatographic condition

Chromatographic separation was carried out at HPLC Shimadzu LC-20 with photodiode array detector (PDA). A 20 μL of sample solution and the standard was injected into the HPLC system. Acetonitrile (ACN) and formic acid 0.1 % was used as mobile in HPLC analysis.

### 2.4. Method Validation

Analytical performance of the method for chromatographic fingerprint analysis was evaluated by determining the precision, repeatability, stability, and specificity. The precision, repeatability, and stability are expressed as relative standard deviation (RSD) of the relative retention time (RRT) and relative peak area (RPA).

## 3. Result and Discussions

### 3.1. Optimization of HPLC Conditions

Quality control methods based on HPLC fingerprints need to be optimized in order to obtain a good separation between the peaks and to give an informative results. In this study, optimization of HPLC condition was conducted through investigating the influence of mobile phase type, mobile phase composition, and column temperature. Acetonitrile (ACN) and formic acid 0.1 % was used as mobile phase at the flow rate of 1 mL / min. Detection was conducted using photodiode array detector (PDA).

Evaluation of optimum condition of HPLC was performed based on number of peak, resolution (R≥1.5), and signal to noise ratio (S/N≥3).

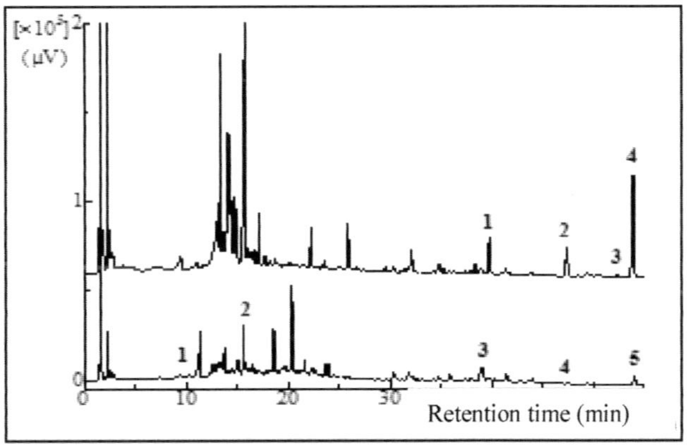


**Figure 1.** HPLC chromatogram of stem (lower) and leaves (upper) of brotowali at optimum HPLC conditions

The optimum chromatographic separation was achieved using linear gradient elution of acetonitrile (ACN) and 0.1% formic acid in water with 5-10% ACN for 0-5 minutes, 10-60% ACN for 5-30 minutes, and 60-95% ACN for 30-55 minutes. Detection wavelength for fingerprint analysis of brotowali was set at 254 nm since most of organic compounds such as aromatic compounds absorbed radiation at this wavelength [8]. Wavelength at 254 nm provided straight baseline and higher sensitivity than other wavelengths. HPLC chromatogram of brotowali stem and leaves at optimum HPLC condition was shown in Figure 1. The optimum of chromatographic conditions produced a peak with the resolution of each peak greater than 1.5 and S/N greater than 3. About 32 peaks for stems and 23 peaks for leaves were detected and could be used as fingerprint of brotowali.

### *3.2. Method Validation*

The precision, repeatability, and stability were expressed as relative standard deviation (RSD) of the relative retention time (RRT) and relative peak area (RPA). The precision was evaluated based on injection precision. The injection precision was determined by five replicate injection of the sample solution in the same day. The repeatability was assessed by analyzing five independently prepared samples of brotowali (stems and leaves). The RSD of RRT and RPA at injection precision test were less than 1.7% and 4.5%, for stems and leaves respectively. Meanwhile, the RSD of RRT and RPA at repeatability were less than 2.6% and 4.6%, for stems and leaves respectively.

The sample stability was determined through injection of a sample solution at 0, 3, 6, and 24 hours after it was prepared. The sample solution was stored in the refrigerator at the temperature of 5-10 °C. The RSD of RRT and RPA in stability test of stems and leaves were less than 5%, indicated that the sample relatively stable for 24 hours storage. Analytical performance of developed HPLC fingerprint analysis of brotowali showed a reliable result, indicated that the proposed method could be used as quality control method of brotowali .

In this study, we also evaluated the specificity of the optimum HPLC fingerprint analysis method. The specificity was evaluated through comparing the HPLC fingerprint profile of brotowali with binahong and green betel leaves. The leaves of those two medicinal plants are almost similar with brotowali leaves. In addition, the dried leaves of each sample were almost indistinguishable in shape and color. The HPLC fingerprint profile of brotowali was clearly distinguishable from HPLC fingerprint profile of those two other leaves (Figure 2). The differences in HPLC fingerprint profile of three samples indicated that the type and content of chemical constituents in each samples were also different. Accordingly, the developed HPLC fingerprint method was specific for brotowali.

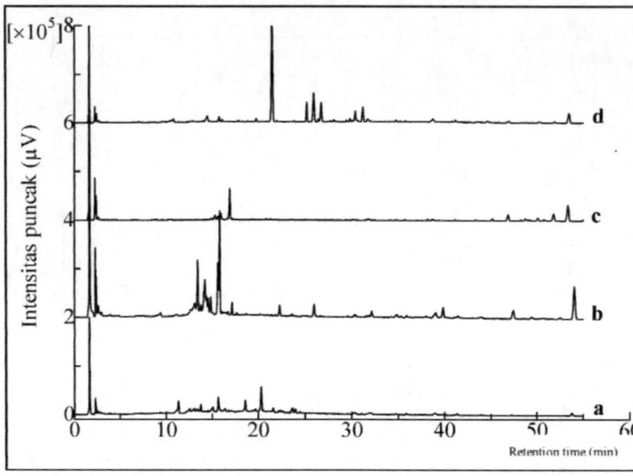


**Figure 2.** HPLC chromatogram of brotowali stem (a) and leaves (b), binahong leaves (c), and green betel leaves (d) using the optimum HPLC condition

### *3.3 Identification of Berberine in Brotowali*

Berberine is one of the important quaternary alkaloids in brotowali (Figure 3). Identification of berberine in stems and leaves of brotowali was conducted by using the developed HPLC conditions. Berberine in sample was identified by comparing the retention time of sample and berberine standard. Berberine in standard solution was indentified at retention time of 20.525 minutes (Figure 4a). Peak at the same retention time was undetectable in stem and leaves extract chromatogram (Figure 4b and 4d). Peaks at retention time of 20.5 were detected in stem and leaves extract after addition (spike) of berberine standard, indicated that berberine was present in very low concentration in both sample [14].

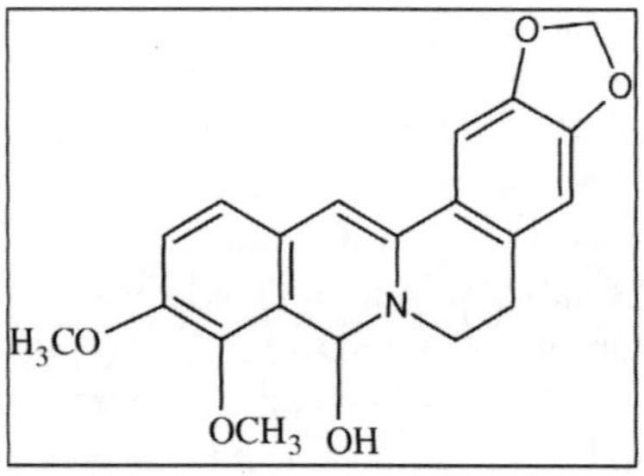


**Figure 3.** Berberine structure

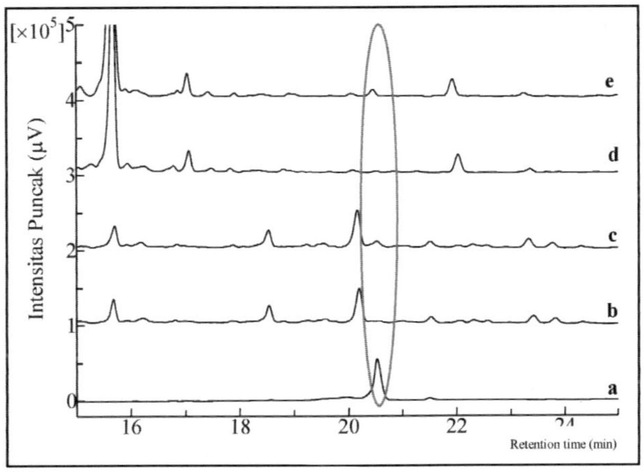


**Figure 4.** HPLC chromatogram of 100 ppm berberine standard (a), brotowali before (b) and after (c) spike with berberine standard, brotowali leaves before (c) and after (d) spike with berberine standard

Berberine was reported as anticancer [15], antimalarials [16], antibacterial, antipyretic, anti-inflammatory [17], and antidiabetic [18]. Although berberine has some pharmacological activity, Indonesian Food and Drug Administrator [19] issued regulations on the prohibition of production and circulation of traditional medicines and health supplements containing berberine. Berberine cause irritation of the kidneys and exhibited nephrotoxic effect.

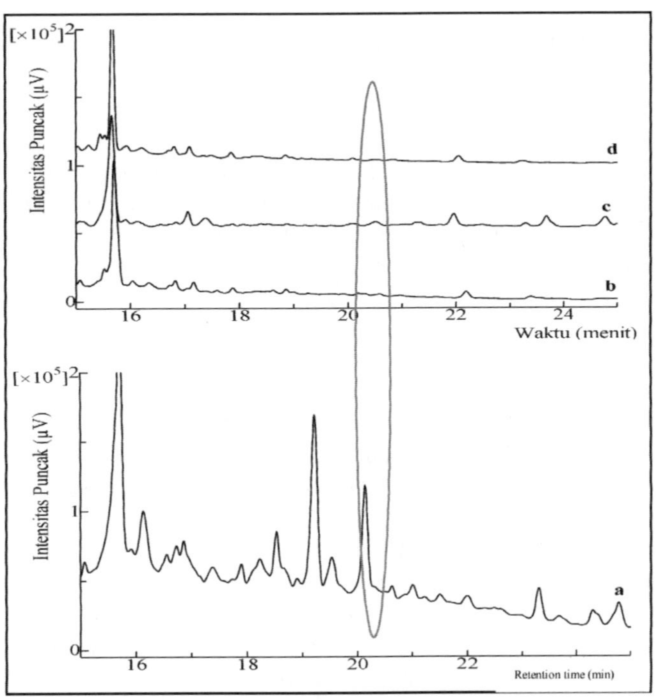


**Figure 5.** HPLC chromatogram of 50% etanol extract of brotowali stem (a), leaves (b), and water extract of brotowali stem (c) and leaves (d)

Berberine which is contained in the brotowali should be eliminated in order to avoid its negative side effect. Choosing the extraction solvent was believed to be one of the most effective effort to eliminate the berberine. In this study, we also evaluated the influence of different extraction solvents on the berberine content in the extracts. Water and 50 % of ethanol were used as extraction solvent of brotowali. Each extract was analyzed using optimum HPLC condition. The 50% ethanol extract showed higher berberine peak than water extract (Figure 5) both for stem and leaves, indicated that water was recommended as extraction solvent for brotowali since the extracted berberine content was lower in water extracts than in 50 % ethanol extracts.

## 4. Conclusion

High performance liquid chromatography fingerprint analysis has already developed for quality control of brotowali. The optimum mobile phase for brotowali fingerprint analysis was acetonitrile (ACN) and 0.1% formic acid in gradient elution with 5-10% ACN for 0-5 minutes, 10-60% ACN for 5-30 minutes, and 60-95% ACN for 30-55 minutes and detection was conducted at wavelength of 254 nm. The analytical performance of developed method met the criteria for parameter of precision, repeatability, stability, and specificity. Berberine as marker compound of brotowali was detected at retention time of 20.525 minutes. Water was recommended as extraction solvent for brotowali in order to minimize the berberine content in the extract.

## References

[1] Hussin MH, Kassim MJ, Razali NN, Dahon NH, and Nasshorudin D 2011 *Arabian J Chem.* **7** 1-9
[2] Hipol RLB, Cariaga MFNM, and Hipol RM 2012 *J Nat Stud.* **11** 88-95
[3] Noor H and Ashcroft S J H 1998 *J. Ethnopharmacol.* **62** 7-13
[4] Ibrahim MJ, I'zzah WN, Narimah AHH, Asyikin N, Shafinas SN, and Froemming GA 2010 *Biomed Res.* **22** 57-62
[5] Rahman NNNA, Furuta T, Kojima S, Takane K, and Mohd MA. 1999 *J Ethnopharmacol.* **64** 249-54
[6] Zakaria ZA, Zakaria ML, Amom Z, and Desa MNM 2011 *Afr J Microbiol Res.* **5** 5379-83
[7] Sulaiman MR, Zakaria ZA, and Lihan R. 2008 *Int J Tropic Sci.* **3** 66-9
[8] Rafi M 2013 Quality control methods for some Zingiberaceous plants from Indonesia using liquid chromatography combined with chemometrics [disertasi]. (Gifu: Gifu University)
[9] Zeng Z, Chau F, Chan H, Cheung C, Lau T, Wei S, Mok DK, Chan C, and Liang Y 2008 *Chin Med.* **3** 9–15
[10] Alaerts G, Matthijs N, Smeyers-Verbeke J, and Heyden YV 2007 *J Chromatogr A.* **1172** 1–8
[11] Zhang YB, Shaw PC , Sze CW, Wang ZT, and Tong Y 2008 *J Food and Drug Anal.* **15** 1-9
[12] Yusoff M, Hamid H, and Houghton P 2014 *Molecules.* **19** 1201-11
[13] Steriti R 2010 *Nat Med J.* **2** 5-6
[14] Skoog DA, West DM, Holler FJ, and Crouch SR 2004 *Fundamental of Analitycal Chemistry Eighth Edition* ( Beltmont: Thomson Learning Inc)
[15] Tang J, Feng Y, Tsao S, Wang N, Curtain R, and Wang Y 2009 *J Ethnopharmacol.* **126** 5-17
[16] Ihwan, Rifa'I M, and Fitri LE. 2014 *J Kedokteran Brawijaya.* **28** 91-96
[17] Mehra R, Naved T, Arora M, and Madan S 2013 *J Adv Pharm Edu and Res.* **3** 440-9
[18] Yin J, Ye J, and Jia W 2012 *Acta Pharm Sin B.* **2** 327-34
[19] [BPOM RI] Badan Pengawasan Obat dan Makanan Republik Indonesia. 2014. Larangan Memproduksi dan Mengedarkan Obat Tradisional dan Suplemen Kesehatan yang Mengandung *Coptis* sp., *Berberis* sp., *Mahonia* sp., *Chelidonium majus*, *Phellodendron* sp., *Arcangelica flava*, *Tinosporae radix*, dan *Cataranthus roseus*. (Jakarta: BPOM RI)

IOP Conf. Series: Journal of Physics: Conf. Series **835** (2017) 012017 doi:10.1088/1742-6596/835/1/012017

# The accumulation of two neolignan in the leaves, stems, and flower of red betel (*Piper crocatum* Ruiz & Pav.)

**Y S Hartini[1*], L Nugroho[2]**

[1]Faculty of Pharmacy, Sanata Dharma University, Yogyakarta, Indonesia
[2]Faculty of Biology, Universitas Gadjah Mada, Yogyakarta, Indonesia

Email : **yustinahartini@usd.ac.id**

**Abstract.** Organ for the biosynthesis of secondary metabolite is not always a place for its biosynthesis, even the same compound was synthesized in the different organs in different plants. Neolignan is a secondary metabolite, known as an imunostimulant, synthesized through shikimic acid pathway with the important precursor is chorismic acid. The compound was known to be accumulated in the roots, stems and leaves of *Piper regnellii* with the concentration varies depending on the type of neolignan. It has been investigated the accumulation of two compound neolignans (Pc-1 and Pc-2) isolated from the methanol extract of red betel leaf (*Piper crocatum* Ruiz & Pav.) in the leaves, stems, and flowers of red betel. Chromatographic methods used was Gas Chromatography-Mass Spectrometry (GC-MS). Chromatogram of GC-MS showed that the Pc-1 with purity of 100%, m/z 460.3 could be detected at the minute of 29.986, while the Pc-2 with purity of 96.681%, m/z 418.3 was detected at the minute of 29.495. The research was then continued to investigate the existence and accumulation of both compounds in leaves, stems, and flowers of red betel. The GC-MS chromatogram shows that Pc-1 and Pc-2 could be detected in the leaves, stem and flower with various concentration among plant organs. Moreover, leaves contained the highest concentration of Pc-1 and Pc-2 compared to other plant organs.

## 1. Introduction

To live, grow, and reproduce, plant transform and interconvert a vast number of organic compounds. The compounds are made by a series of reaction called metabolism which could be distinguished to be primary and secondary metabolisms. Humans utilize secondary metabolites for various purposes because the compounds show various pharmacological activities. Secondary metabolites were synthesized in certain organs of the plant. Plant secondary metabolites were accumulated in different ways. Tobacco, for instance, synthesize its alkaloid nicotine in the root but accumulate it in the leaves [1], while *Piper regnellii* accumulate its neolignan in the roots, stems, and leaves [2]. *Piper crocatum* known as red betel is a species of the family Piperaceae. Decoction of the leaves of the plant family Piperaceae used traditionally for the treatment of various diseases. Our previous study reported that two neolignans isolated from methanolic extract of *P. crocatum* leaves had imunostimulant actitivity [3]. Regarding many previous studies, the accumulation of both compounds could be in the organs other than leaves. In order to broaden the provided material for traditional healing, the distribution of both compounds in organs other than leaves should be known. After knowing the distribution of both

* Corresponding author

compounds, hopefully the provided material for traditional healing could also be supplied by organs other than leaves. Therefore, the accumulation of two compound neolignans (Pc-1/pipercrocatin and Pc-2/deacetyl pipercrocatin) isolated from the methanolic extract of red betel leaf in the leaves, stems, and flowers were detected in this study.

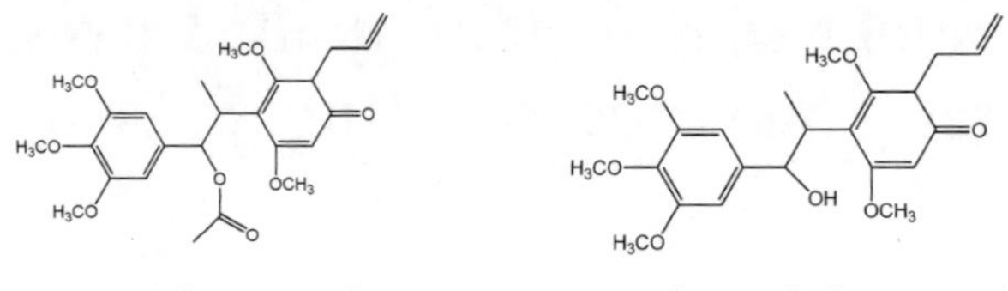


**Figure 1.** Structures of the neolignans isolated from *P. crocatum* [4,3]

## 2. Experimental

### *2.1. Plant material*
The leaves, steam, and flowers of red betel were collected in April 2015 at Sleman, Yogyakarta Indonesia. The plant material was identified by Wahyono at the Department of Pharmaceutical Biology, Faculty of Pharmacy, Universitas Gadjah Mada, Yogyakarta, Indonesia and a voucer specimen (no.BF/284/Ident/Det/VIII/2011) was deposited in herbarium unit at The Faculty of Pharmacy, Sanata Dharma University, Yogyakarta, Indonesia.

### *2.2. Extract preparation*
The leaves, stems and flowers of red betel were dried and powdered using mortar and pestle. The powder of three organs were then extracted with methanol by maceration method at a room temperature for three days. The extract were filtered and evaporated under vacuum.

### *2.3. Sample preparation*
Stock solution of Pc-1, Pc-2, and the extracts of the leaves, stems and flowers from red betel were prepared in the mixture of chloroform:methanol (1:1).

### *2.4. Instrumentation and chromatographic conditions*
The analysis were carried out using an Agilent GC 6890N 59765 B MSD connected with MS. The capillary column was Agilent 19091S-433 HP-5ms 5% phenyl methyl siloxane. Maximal temperature was 325°C with the nominal length of 30,0 m, the diameter was 250 μm, the film thickness was 0.25 μm, the initial flow was 1.0 mL/min, the nominal initial pressure was 8.65 psi, and the average velocity was 37 cm/sec.

## 3. Result and Discussion
Pipercrocatin (Pc-1) compound was identified in the form of a white crystalline glazed, while the form of deacetyl pipercrocatin (Pc-2) was greenish-white coloured powder. The Figure 2-4 shows the GC-MS chromatograms of Pc-1 and Pc-2, the MS spectrograms of Pc-1 and Pc-2, and the GC-MS chromatogram of red betel leaves methanolic extract.

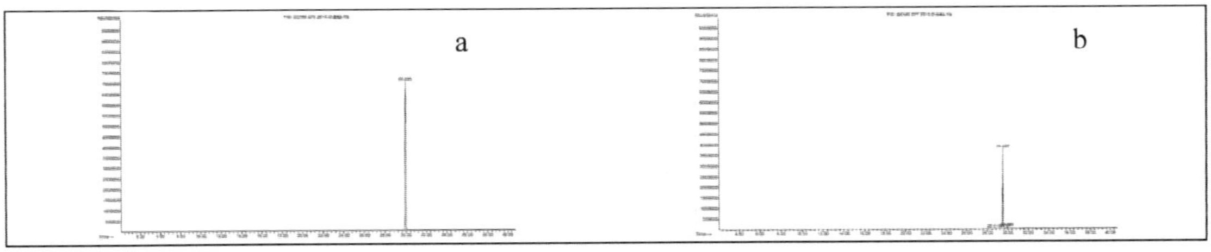


**Figure 2.** The GC-MS chromatogram of Pc-1 (a) and Pc-2 (b)

The maximum absorption of Pc-1 and Pc-2 were found at 234,5 nm and 264,5 nm 280 nm, respectively. Chromatogram of GC-MS showed that the Pc-1 with the purity of 100%, m/z 460.3 could be detected at the minute of 29.986, while the Pc-2 with the purity of 96. 681%, m/z 418.3 was detected at the minute of 29. 495.

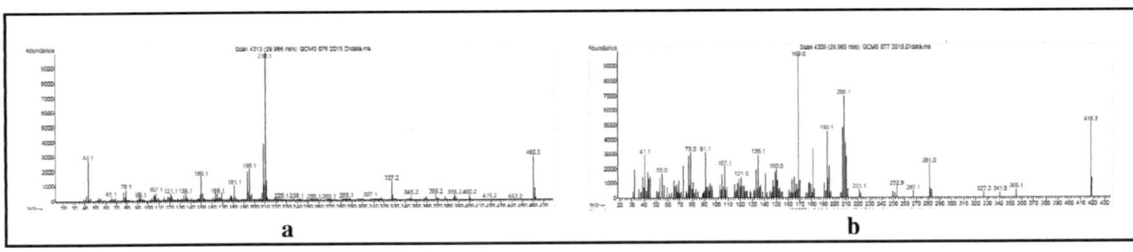


**Figure 3.** The MS spectrogram of Pc-1 (a) and Pc-2 (b

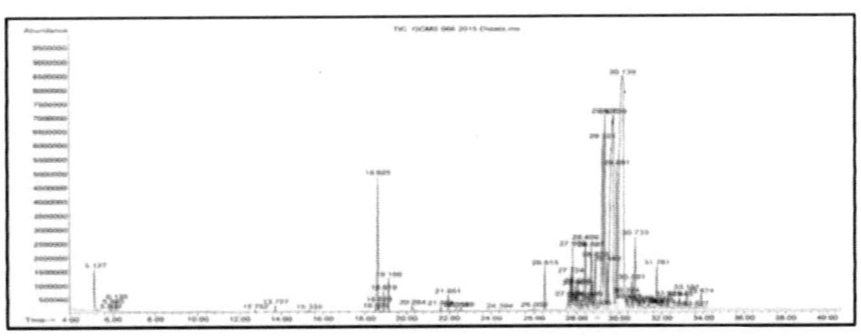

**Figure 4.** The GC chromatogram of red betel leaves extract

**Table 1**. Area percent of Pc-1 and Pc-2 in the leavess, stem, and flowers of red betel.

| Red betle extract of: | Pc-1 | Pc-2 |
|---|---|---|
| Leaves | 31.726% | 19.798 % |
| Stems | 29.531 % | 13.533 % |
| Flowers | 5.259 % | 16.209 % |

Organ for the biosynthesis of secondary metabolite is not always a place for its biosynthesis, even the same compound was synthesized in the different organs in different plants. The extracts demonstrate similar chromatographic profile, but there were differences in the concentrations of neolignans. Neolignan is a secondary metabolite, synthesized through shikimic acid pathway. Some neolignans participate in lignin synthesis and hence have important roles in determining physicochemical and mechanical properties of cell walls, whereas others act as antioxidants, biocides (fungicides, bactericides and antiviral agents) and perhaps even as cytokinins [5]. Our previous research found that Pc-1 and Pc-2 showed imunostimulant activity [3]. Table 1 shows the content of neolignans in the different parts of the plant. Leaves presented a highest concentration of Pc-1 and Pc-2. The lowest concentration of Pc-1 was found in flowers, while stems contain the lowest concentration of Pc-2. The reason why leaves contain the highest concentration of Pc1 and Pc2 may due to the important precursor for various compounds in the shikimic acid pathway, namely chorismic acid was located in the plastids [6]. Therefore, the accumulation of most chorismic acid derivates including lignan were abundance in the organ with higher content of plastid, its means was leaves. Differences in the distribution of neolignan compounds are also shown in the other Piper species namely *P. regnellii*. The compound was known to be accumulated in the roots, stems and leaves of *P. regnellii* with the concentration varies depending on the type of neolignan [2]. The results of this study recommend the use of organ leaves to get neolignan compound especially Pc-1 and Pc-2.

## 4. Conclusion

The neolignans (Pc-1 and Pc-2) isolated from red betel (*Piper crocatum* Ruiz & Pav.) accumulated in the leaves, stems, and flowers of red betel with the highest concentrations in the leaves.

## References

[1] Shoji T, Kajikawa M and Hashimoto T 2010 *Plant. Cell* **22** 3390-409
[2] Felipe D F, Filho B P D, Nakamura C V, Franco S L and Cortez DAG 2006 *J. Pharm. Biomed. Anal.* **41** 1371-5
[3] Hartini Y S, Wahyuono S, Widyarini S, Yuswanto A 2014 *Trop. J. Pharm. Res.* **13** 1609-14
[4] Kustiawan P M 2012 *Isolasi dan Identifikasi Senyawa Imunostimulan Non Spesifik In Vitro dari Daun Sirih Merah (Piper crocatum Ruiz & Pav.)* (Yogyakarta: Fakultas Farmasi Universitas Gadjah Mada) p 65
[5] Lewis N G and Davin L B 1994 *Evolution in lignan and neolignan biochemical pathways, in Evolution of Natural Proucts* (Washington DC : ACS Symposium series) pp. 202-45
[6] Verberne M C, Verpoorte R, Bol J F, Mercado-Blanco J and Linthorst H J M 2000 *Nat. Biotechnol.* **18** 779-83

## Acknowledgment

The author is grateful to Ministry of Research, Technology and Higher Education, Republik of Indonesia for the financial support by Fundamental Research Grant with contract References Grant with contract number: 010/HB-LIT/III/2015.

# High performance thin layer chromatography fingerprint analysis of guava (*Psidium guajava*) leaves

**M Astuti**[1], **L K Darusman**[1,2], **M Rafi**[1,2*]

[1]Department of Chemistry, Faculty of Mathematics and Natural Sciences, Bogor Agricultural University, Dramaga, Bogor, West Java Indonesia
[2]Tropical Biopharmaca Research Center, Bogor Agricultural University, Taman Kencana, Bogor, West Java Indonesia

E-mail: mra@ipb.ac.id

**Abstract**. High-performance thin layer chromatography (HPTLC) fingerprint analysis is commonly used for quality control of medicinal plants in term of identification and authentication. In this study, we have been developed HPTLC fingerprint analysis for identification of guava (*Psidium guajava*) leaves raw material. A mixture of chloroform, acetone, and formic acid in the ratio 10:2:1 was used as the optimum mobile phase in HPTLC silica plate and with 13 bands were detected. As reference marker we chose gallic acid ($R_f$ = 0.21) and catechin ($R_f$ = 0.11). The two compound were detected as pale black bands at 366 nm after derivatization with sulfuric acid 10% v/v (in methanol) reagent. Validation of the method was met within validation criteria, so the developed method could be used for quality control of guava leaves.

## 1. Introduction

Guava (*Psidium guajava*) leaves known medicinal plant especially for diarrhea treatment and widely distributed in Indonesia. Guava leaves are commonly used in Indonesia to treat a cough, toothache, digestive problems, skin problems, laryngitis, anorexia, rheumatic, epilepsy, diabetes, hypertension, etc. [1]. A chemical compound present in plants could have different in the concentration because affected by several factors, such as differences in geographical origin, time of harvest, climate, drying processes, and other factors [2].

The efficacy of medicinal plant is affected by its chemical composition and concentration, so identification of medicinal plant is needed to prevent any chance of substitution or to mix with another plant. As we know, the quality evaluation of the medicinal plants usually based on the detection or quantitation of some active compound or reference marker. Unfortunately, the presence of reference marker did not always guarantee we used the exact medicinal plant. Substitution or mixing with other closely-related medicinal plants could be occurs. Another approach to the development of quality control method of the medicinal plant by using fingerprint analysis [3,4].

Fingerprint analysis is an effective method to control and evaluate the quality of raw materials and herbal medicinal product. This approach is also able to monitor the quality of the medicinal plants because able to show the overall composition of the chemical components and their relative concentrations. HPTLC fingerprint analysis becomes the most widely used for quality control of

[*] Corresponding author

medicinal plants and their related products because of its simplicity and reliability. Also, the result from this method could be directly evaluated by visual inspection [5]. The main application of HPTLC fingerprint analysis is for identification and authentication of medicinal plants [6-8]. In this work, we have been developed an HPTLC fingerprint analysis for identification and authentication of guava leaves that could be used as a quality control method of its raw material.

## 2. Materials and Methods

### *2.1 Materials*
Guava leaves were obtained from Cimanggu and Dramaga, Bogor and also from Conservation and Cultivation Unit of Tropical Biopharmaca Research Center, Bogor Agricultural University, Indonesia. Gallic acid (≥97.5%) and (+)-catechin (99.0%) were purchased from Sigma-Aldrich (Missouri, USA). HPTLC silica gel 60 $F_{254}$ methanol, formic acid, acetone, chloroform, and sulfuric acid were obtained from Merck (Darmstadt, Germany).

### *2.2 Sample and standards preparation*
Dried powder of guava leaves (1 g) was mixed with 10 mL methanol and extracted in an ultrasonic bath for 30 minutes at room temperature, and then the extract was filtered. Gallic acid and catechin solutions were prepared by dissolving 1.25 mg of standards in 5 methanol to obtain concentration about 250 μg/mL.

### *2.3 HPTLC fingerprint analysis*
Methanol extract of guava leaves (5 μL) and standards solution (2.5 μL) were applied to HPTLC plate as 8 mm bands, at least 4 mm apart, 10 mm from lower edge and at least 15 mm from left and right edge of the plate using TLC semiautomatic sampler Linomat 5 (CAMAG, Muttenz, Switzerland). The plate was developed to a distance of 8 cm in a 20 cm × 10 cm twin trough chamber (CAMAG, Muttenz, Switzerland). As the optimum mobile phase, we used a mixture of chloroform, acetone, and formic acid in the ratio 10:2:1.

### *2.4 Derivatization and evaluation*
Sulfuric acid 10% (v/v) reagent was prepared by mixing 10 mL sulfuric acid with an ice-cooled of 90 mL methanol. Documentation of developed plate used CAMAG Reprostar 3 (CAMAG, Muttenz, Switzerland) with WinCATS software. Before derivatization, plate was documented and evaluated under UV light at 366 nm. For derivatization, the plate sprayed with reagent then heated at 100 oC for 5 minutes. Derivatized plate was also documented and evaluated under UV 366 nm.

### *2.5 Method validation*
Validation of the HPTLC fingerprint analysis of guava leaves was evaluated following the guidelines described by Reich and Schibli [5] by determining the stability of analyte on the plate, in solution, and during chomatography, the stability of derivatized zones, specificity, repeatability, intermediate precision, and robustness.

## 3. Results and Discussions

### *3.1 Method Development*

3.1.1 *Optimization of mobile phase composition.* Optimization of the mobile phase for HPTLC fingerprint analysis of guava leaves as an initial step in the method development began using ten single solvents with different degree of polarity. A mixture of chloroform and acetone was chosen because they gave good separation with the highest number of bands compared to other solvents (data not shown). We added formic acid in the mixture of chloroform and acetone to increase the polarity of the mobile phase to increase the $R_f$ of marker compounds (gallic acid and catechin). The optimum mobile phase composition was found in the ratio 10:2:1 of chloroform, acetone, and formic acid respectively.

Figure 1 showed the HPTLC fingerprint chromatogram using the optimum mobile phase with detection under 366 nm UV light after sprayed with sulfuric acid 10%. Gallic acid and catechin were separated well and also we obtained another 13 bands in the fingerprint chromatogram. The $R_f$ of gallic acid was 0.21 while catechin was 0.11 and both bands detected as pale black bands.

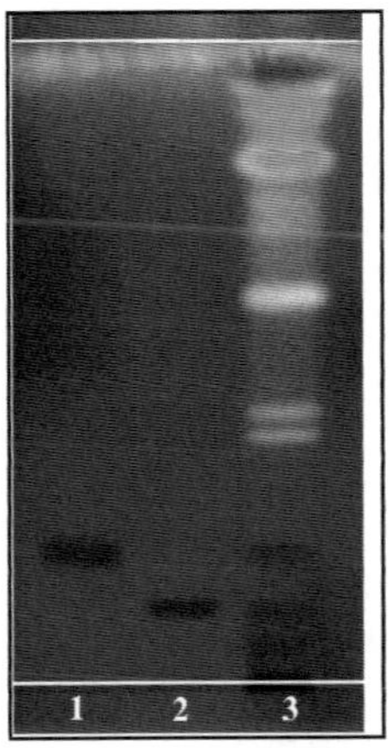


**Figure 1.** HPTLC chromatogram of guava leaves extract. Track 1: gallic acid, 2: catechin, 3: guava leaves methanol extract

*Validation of the developed method*

The developed method was validated by determining the stability of analyte on the plate, in solution, and during chromatography, the stability of derivatized zones, specificity, repeatability, intermediate precision, and robustness. The focus on validation of the developed HPTLC qualitative method for identification of guava leaves was fingerprint profile, such as $R_f$ value and color bands. The results obtained from the validation step are evaluated according to criteria described by Schibli and Reich [5].

3.1.2 *Stability.* Stability of analyte on the plate, in solution, and during chromatography were evaluated, and the results should not be different in the position and color of bands. Based on the result obtained in Figure 2, the guava leaves extract is stable on the HPTLC plate and in solution because no differences are seen in zone intensity, and there are no zones that appear or disappear. Therefore, the sample is considered stable for at least 3 hours in solution and on the plate.

Stability of the sample extract during chromatography was investigated by 2-D development. If the sample is stable during chromatography, all components should be detected on the diagonal line connecting the application position and the intersection of the two solvent front. Figure 3 showed the 2-D HPTLC chromatogram, and the sample is stable during chromatography because no zones are located off of the diagonal. Therefore, the sample is considered stable during chromatography.

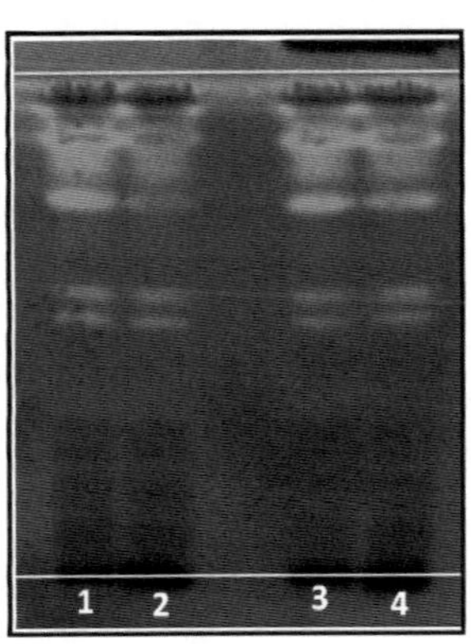


**Figure 2.** HPTLC chromatogram which showed stability on plate and in solution. Track: 1: first extract applied on plate for 3 hours prior to chromatography, 2 and 4: second extract applied immediately prior to chromatography, 3: first extract 3 hours in solution and applied immediately prior to chromatography

IOP Conf. Series: Journal of Physics: Conf. Series **835** (2017) 012018 doi:10.1088/1742-6596/835/1/012018

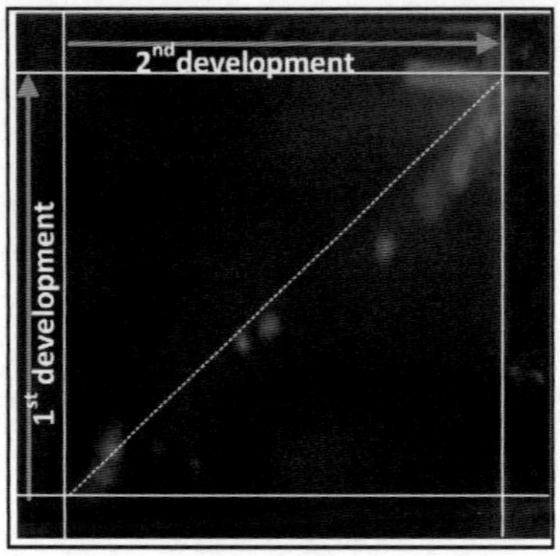


**Figure 3.** HPTLC two-dimensional chromatogram of guava leaves extract which showed stability during chromatography.

In the stability of derivatization, if there is no significant change in the image within 30 min meaning the derivatization yields are stable. As can be seen in Figure 4, after derivatization, the chromatogram is stable for at least 1 hours.

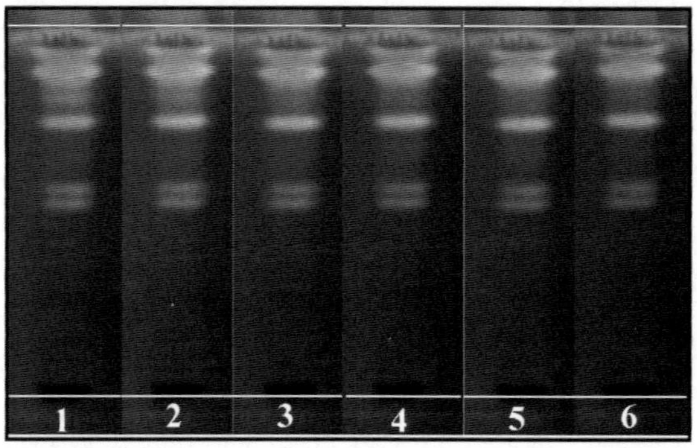


**Figure 4.** HPTLC chromatogram of guava leaves extract stable after derivatization at least 1 hours. Track: 1: after 2 minutes, 2: after 5 minutes, 3: after 10 minutes, 4: after 20 minutes, 5: after 30 minutes, 6: after 60 minutes

*3.1.3 Specificity.* The number, color, intensity, and position of the zones in the HPTLC fingerprint chromatogram was evaluated for the specifity. A method will be called specific if the sample tested from different location growth gives similar fingerprint profile, and the reference marker is also detected. The developed method provides similar pattern in the HPTLC chromatogram of guava leaves from different location growth only differ in the intensity of the detected bands, but all of the samples contained gallic acid and catechin as the reference marker (Figure 5).

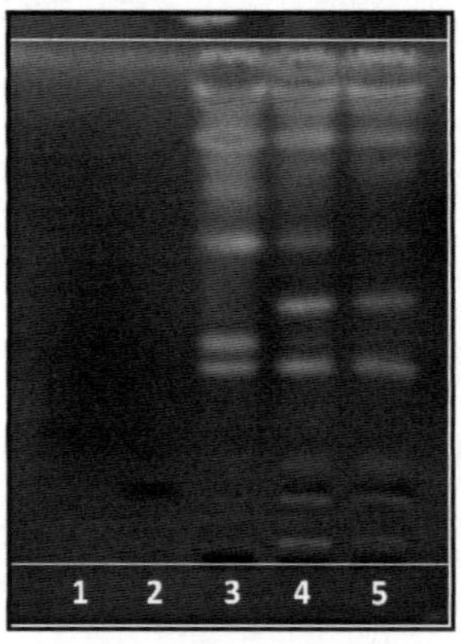


**Figure 5.** The HPTLC fingerprint profiles of various guava leaves extract from different location growth. Track: 1: gallic acid, 2: catechin, 3: dried guava leaves from Tropical Biopharmaca Research Center, 4: fresh guava leaves from Cimanggu, Bogor 5: fresh guava leaves from Dramaga, Bogor.

*3.1.4 Precision.* Evaluation in the precision (same day) and intermediate precision (three consecutive days) of the HPTLC fingerprint analysis of guava leaves is determined by the repeatability of the selected bands $R_f$ values for each of the three zones on the three plates. We chose gallic acid, catechin and one compound at the $R_f$ 0.75 as the selected bands for the determination of the developed method precision. Variation in the $R_f$ values from the three selected bands are less or equal than 0.02 and 0.05 respectively (Table 1), so the developed method is precise.

**Table 1.** Precision (same day) and intermediate precission (three consecutive day) for three selected zones on HPTLC fingerprint analysis of guava leaves

| Compounds | $R_f$(repetition) | | | $\Delta R_f$ |
|---|---|---|---|---|
| | $1^{st}$ | $2^{nd}$ | $3^{rd}$ | |
| Precision | | | | |
| Catechin | 0.13 | 0.12 | 0.11 | 0.02 |
| Gallic acid | 0.21 | 0.20 | 0.19 | 0.02 |
| *X* compound | 0.75 | 0.73 | 0.73 | 0.02 |
| Intermediate precision | | | | |
| Catechin | 0.14 | 0.13 | 0.13 | 0.01 |
| Gallic acid | 0.21 | 0.20 | 0.19 | 0.02 |
| *X* compound | 0.81 | 0.81 | 0.76 | 0.05 |

*3.1.5 Robustness.* In this study, we used two criteria for evaluation of the robustness. The $R_f$ values of the reference marker (gallic acid and catechin) bands must lie within the acceptance criteria with differences in the $R_f \leq 0.05$ when we used two type of chamber (twin through and flat bottom) and two different lengths (7 and 8 cm) for developing distance. From the results obtained in Table 2, the differences of the $R_f$ values is more than 0.05 when we used different chamber type; it means the flat-bottom chamber gives different results so this method should only perform using twin trough chamber. Decreasing in the developing distance from 8 cm to 7 cm did not affect so much in the $R_f$ values of the two reference marker (Table 3).

**Table 2.** $R_f$ values from reference marker zones on different chamber type

| Compounds | $R_f$(chamber type) | | $\Delta R_f$ |
|---|---|---|---|
| | Twin-trough | Flat-bottom | |
| Catechin | 0.11 | 0.20 | 0.09 |
| Gallic acid | 0.21 | 0.35 | 0.14 |

**Table 3.** $R_f$ values from reference marker zones on different developing distance

| Compounds | $R_f$(developing distance) | | $\Delta R_f$ |
|---|---|---|---|
| | 8 cm | 7 cm | |
| Catechin | 0.11 | 0.12 | 0.01 |
| Gallic acid | 0.21 | 0.22 | 0.01 |

## 4. Conclusion

The optimum mobile phase for HPTLC fingerprint analysis of guava leaves extract has used a mixture of chloroform, acetone, and formic acid in the ratio 10:2:1. Gallic acid and catechin were used as reference marker with the $R_f$ 0.21 and 0.11, respectively. The developed validated HPTLC fingerprint analysis method could be used as identification and authentication method of guava leaves.

## References

[1] Gutierrez RMP, Mitchell S and Solis RV 2008 *J. Ethnopharmacol.* **117** 1-27
[2] Liang YZ, Xie P and Chan K 2004 *J. Chromatogr. B* **814** 53-70.
[3] Soares PK, Burns RE, Almeida AA and Scarminio IS *J. Sep. Sci.* **30** 3302-10.
[4] Rafi M, Wulansari L, Heryanto R, Darusman LK, Lim LW and Takeuchi T 2015 *Food Anal. Methods* **8** 2185–93.
[5] Reich E and Schibli A 2006 *High-Performance Thin Layer Chromatography for the Analysis of Medicinal Plants* (New York: Thieme Medical Publishers)
[6] Gunalan G, Saraswathy A and Vijayalakshmi K 2012 *Asian Pac. J. Trop. Dis.* **2** S21-S25.
[7] Sudberg S, Sudberg EM, Terrazas J, Sudberg S, Patel K, Pineda J and Fine B 2010 *J. AOAC Int.* **93** 1367-75
[8] Reich E, Schibli A and De Batt A 2008 *J. AOAC Int.* **91** 16-8

**Institute of Physics**
Dirac House, Temple Back
Bristol BS1 6BE UK

ISSN: 1742-6588
ISBN 978-1-5108-4077-5

9 781510 840775

CURRAN ASSOCIATES INC.
proceedings.com

# International Interdisciplinary Conference "Euler Readings MRSU 2017"

Journal of Physics: Conference Series Volume 996

Moscow, Russia
22 – 24 November 2017

IOP
Institute of Physics